RESTORING HOUSES OF BRICK & STONE

RESTORING HOUSES OF BRICK & STONE

NIGEL HUTCHINS
and
Donna Farron Hutchins

KEY PORTER BOOKS

Published in hardcover in 1982 by Van Nostrand Reinhold Publishers

This edition published in 1998 by Key Porter Books

Canadian Cataloguing in Publication Data

Hutchins, Nigel, 1945–
Restoring houses of brick and stone

Rev. ed.
Includes bibliographical references and index.
ISBN 1-55013-962-2

1. Dwellings—Remodeling. 2. Brick houses—Conservation and restoration.
3. Stone houses—Conservation and restoration. I. Title

TH4816.H87 1998 643'.7 C97-932269-3

The publisher gratefully acknowledges the support of the Canada Council for the Arts and the Ontario Arts Council for its publishing program.

Key Porter Books Limited
70 The Esplanade
Toronto, Ontario
Canada M5E 1R2

Distributed in the United States by Firefly Books

Printed and bound in Canada

98 99 00 01 02 6 5 4 3 2 1

FRONT COVER PHOTOGRAPH: Danny Burke
BACK COVER PHOTOGRAPH: Martin Weaver
COVER DESIGN: Jean Lightfoot Peters
EDITORIAL CONSULTANT: James T. Wills
DESIGN: Brant Cowie/Artplus Ltd.
ILLUSTRATIONS: Graham Thomas
TYPESETTING: Compeer Typographic Services Ltd.

Photography Credits

All photographs are by the author with the exception of the following: Ross Anderson, 59 (above left); Gaye Bennett, 26, 32, 50, 59 (below left and right), 74 (above), 83, 92 (right), 95, 109, 145; Keith Blades, 12, 69 (above), 136; Danny Burke, 134, 161, 163 (below); Richard Byrne, 59 (above right), 139 (below), 167; Milton Droeski, 11, 16, 20; Phil Gerrard, 114; Arthur Holbrook, 61 (above left), 139 (above); Tony Jenkins, 106, 107, 140, 141; Jeremy Jones, 164; The *Ottawa Citizen*, 101, 123, 130; M. Poitras, 181; K. Beth Potter, 131; Public Archives Canada, 22 (C-31743), 25 below (C-31747), 33 above (PA-9734), 39 (C-37501), 45 (PA-40068), 48 (PA-40067, PA-45228), 49 (PA-40050, PA-73722), 51 (PA-51315), 53 (PA-123890), 54 (PA-89867), 57 (C-31784), 79 (PA-86183), 105 (PA-36867); James T. Wills, 96, 137, 143

Contents

Foreword

In *Restoring Houses of Brick & Stone*, Nigel Hutchins has created a worthy successor to *Restoring Old Houses*.

When Nigel Hutchins came to me and said that he and his publisher would again like to enlist the assistance of myself and my staff at the Heritage Canada Foundation in the production of a book, I was intrigued. I wondered just how the new work could continue in the same rich vein as its predecessor, explaining and introducing in simple terms the history, mysteries and complexities of the mason's trade and its products in North America.

The reader will find in the following pages just how successful the latest venture has been.

To illustrate the text, Nigel Hutchins has culled from our collections and those of many others, both public and private, many photographs that are appearing for the first time in print. In some instances, we have had to search long and hard through obscure technical books from the turn of the century to confirm the correct meaning and spelling for technical terms. Heated argument and strong controversy have raged over such subjects as the name of a hammer, but good humor and accuracy have prevailed. I feel that all the work has been well worthwhile because it makes useful information easily accessible to an even wider audience.

One of the fundamental goals of the Heritage Canada Foundation is to promote the management of the inevitable changes in our built environment in such a way that we preserve the character, craftsmanship and beauty of our older buildings for ourselves and our descendants. We feel that old brick and stone homes have a very special place in our built environment.

Nigel Hutchins helps us in our work and opens a door on this special heritage, inviting the reader to come through and participate in its preservation.

Martin E. Weaver
The Heritage Canada Foundation, Ottawa

Preface

In the enjoyable pages that follow, Nigel Hutchins invites the reader into a world of the most rewarding kind of activity, but cautiously so. He reminds us that the always important precursor to doing is knowing.

This splendidly thorough work is at once a comprehensive history and a practical manual. Its starting point is an exhortation to the well-intentioned old-home preservationist to know first and foremost the meaning of the term "restore." For the restorer, historical accuracy is the watchword, even if the result shatters our preferred fancies. For the renovator, there are more options. We may very well wish to rebuild a historic home according to our wishes for what history should have been, or we may take the earnest path and seek to return a structure to what it once was in actual fact—the important thing is that we are aware that we are making a choice.

A great wealth of pleasure accrues from a slow and sensitive reading of each chapter. The historical introductory comments are a special treat, valuable if for no other reason than to spare us the exasperation of repeating the mistakes of those who have gone before. The historical erudition of these pages is matched by a wealth of sensible day-to-day insights regarding selection and repair of building materials, techniques of construction and reconstruction, methods of cleaning, principles of renovation and additions and invaluable discussions of things best not done at all.

For the professional and amateur alike, here is a clear-headed, wise and even witty account of a thousand useful topics. And when you've finished with your renovation or restoration, you can say with pride that you've done it by the book.

Michael Bird
Waterloo, Ontario

Introduction

THE PHYSICAL elements that make up a house are usually those which are readily available or in close proximity to the building being constructed. Today, the local lumberyard often provides a wide selection of wood, brick, stone and other non-traditional materials for use in domestic construction. Similarly, early settlers in North America made full use of the materials that were available to them. The reasons were obvious: transportation for bulky materials was virtually non-existent and, if materials on hand filled the need, then they were used. Wood structures formed one school of material construction, masonry structures another. The preservation of the latter is the subject of this book.

Our architectural inventory may be called our architectural heritage. Just as the first owner-builder was responsible for the creation of the house, the responsibility for its preservation frequently falls on the current owner. Preserving our architectural heritage is an important yet enjoyable task. As the acquisition of older homes becomes more widespread, there is a growing need for information to help homeowners restore their homes to their former elegance and beauty. This book will, I hope, assist the participant in this mission.

There are many people involved in the creation of a book. My associates and colleagues have assisted greatly. I would especially like to thank Stan and Jean Slivinski and my mother, Winnifred Hutchins. I would further like to thank Leslie and Betty Hutchins for their confidence and support. My thanks to Glen Billings, and to Bob and Edith Lenz (Sawmill Antiques); to Martin Weaver and Richard Byrne, who perused the text, commented, offered suggestions—not always in the most diplomatic English—and generally shared their unquestionable expertise with the author; to my associates on this book, Gaye Bennett and Graham Thomas, whose photography and illustrations speak for their obvious talent; to my editor Jim Wills, who often must have felt like murdering me during this project (his skill and notable talents in his own restoration work have been an asset to this text); to the owners of so many houses for the intrusion into their living space; and, lastly, to my associate on this book, Donna Hutchins, without whose talent, stamina and skill I could not have completed this work. To our sons Jesse and Dylan: thank you for your patience and understanding.

CHAPTER ONE

From the Old Land to the New

This timber-frame structure with its rubblestone or pierrotage *infill is a medieval form transported from France. Other variations employed wattle and daub or brick as infill. Louisbourg, Cape Breton Island, early eighteenth century.*

THE PRESERVATION, repair and adaptation of historic North American buildings so that they may be lived in today but kept for tomorrow is the subject of this book. Its underlying theme, however, is an attempt to convey a sympathetic understanding of the period in which these buildings were constructed. A grasp of historical developments is significant for the preservationist, simply because without it aesthetic judgments that affect a structure become mere guesses rather than certainties.

The use of brick and stone in North American domestic architecture has a surprisingly long and venerable history. Although a Georgian stone farmhouse may be a vernacular expression of a style imported from eighteenth-century England, the methods and materials employed are far more ancient. The Georgian style was itself based upon Roman and ultimately Greek models, but construction with brick and stone was known and successfully practised even earlier.

Domestic uses of masonry units can be traced as far back as 3200 B.C. The ordinary Egyptian house of this period was built with a type of adobe brick made up of sand, clay and water, often with straw as the binding agent. This kind of brick was also used along with stone in early Egyptian tombs, called mastabas. The stone was quarried using copper tools and wooden wedges, moved to the building site by means of wedges and levers, and finally drawn into position by teams of men. The most fascinating aspect of this herculean exercise is the fact that the wheel was never used.

Even more difficult for the modern mind to grasp is that these same methods were employed in the building of the Great Pyramid at Giza. More than six million tons of stone were put in place; the joints between them measured no more than one-fiftieth of an inch. Anyone who has ever attempted their own masonry work will justifiably marvel at such a feat. Although the facade of the Giza tomb was faced with exactly cut stone, perhaps limestone from the

This medieval arch was constructed with rubblestone and brick.

Classical column styles

quarries at Tura, south of Cairo, the interior was a solid rubble core made up of rubblestone and adobe brick.

A parallel development, though smaller in scale and later in time, was the post-classical Mayan buildings of 1200 A.D. Once again a rubblestone and adobe core was clad with cut stone in the more important buildings. Mayan artisans were kept busy incising the stone with mythological figures. Domestic buildings, on the other hand, were built almost entirely of adobe brick.

Mayan influence on western culture is minimal, however. The forms and methods of ancient Egypt laid the groundwork for the use of masonry in the western world, while early Greek and Roman cultures spawned the concepts and architectural styles that are still visible in the North American landscape. The abundance of good building stone in Greece gave rise to its appearance in domestic structures. Walling was formed by using rectangular, polygonal or large, undressed (cyclopean) stone. Occasionally, inclined blocks formed the lintels of openings. Sun-dried brick was also employed, but mortar was not a part of Grecian construction methods. In cut stone buildings, iron clamps held the units in position, and hot lead was poured on as a locking device. It is an interesting historical note that in the area around Albany, New York, during the nineteenth century, a variant of this method was used in fireplace construction.

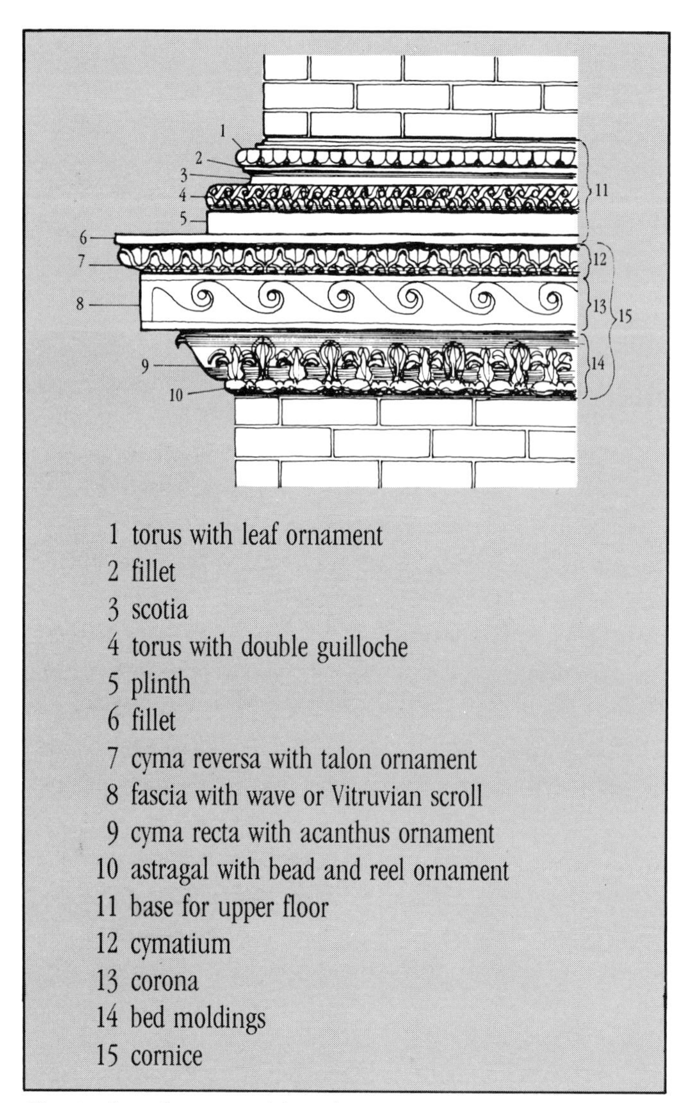

Classical architectural banding

The Greeks attached great importance to the use of marble, which was readily available and which could be worked into smooth surfaces where exactness of form was necessary. Lime stucco was developed to duplicate the solid material. It was commonly used and perhaps formed the basis for the appearance of stucco, ashlar facades in eighteenth- and nineteenth-century North America.

The Greek was the complete artist, combining cleanness of form and spirituality of architecture. By contrast, the Roman was the engineer, garish in style but joyously simple in undertakings of a utilitarian nature. Roman architecture utilized many types of materials for every aspect of the mason's craft. Stone and marble were certainly employed, but so were terra cotta and brick in great quantities. Volcanic earth, called *pozzo lana*, abundant around Rome and Naples, became the basic ingredient of a non-clay mortar. Brick was sun-dried until a simple method for brick firing was found. The Romans improved on this method. The common Roman brick, much stronger than its sun-dried predecessors, was usually a one-inch-thick square similar in appearance to a tile. These bricks were used in many European areas that were under the sway of the Roman Empire. Perhaps this is why the modern Danish word for brick is *tegisten,* which translates into "tile stone."

Greek and Roman influences were not alone in shaping European architectural tradition. Islam was a mighty force, not only in things military but also in things architectural. No clearer statement of Islamic influence in our recent past can be made than that of Sir Christopher Wren in his 1713 *History of Westminster Abbey*:

> This we now call the Gothick manner of architecture ... tho' the Goths were rather destroyers than builders, I think it should with more reason be called the Saracen style; for those people wanted neither arts nor learning.[1]

The rise of Islam can be documented from the seventh century, A.D. The countries drawn into its sphere of influence throughout the next 700 years added stone construction and sophisticated uses of ceramics to an already established tradition of brickmaking. What timber-frame construction there was incorporated the brick and rubble infill that appears in all cultures skilled in this style of building. The use of brick and stone, both decoratively and in their more important structural functions, was the highpoint of Islamic methods. Carved stuccos, stone inlay, ceramic mosaics and banded masonry with its alternating brick and stone courses were all highly refined skills of the Islamic mason.

Despite Roman occupation and Moorish coastal raids, the English do not appear to have commonly employed masonry for domestic architecture before about the twelfth century, although some churches and early rubblestone forms in Ireland do date back to the sixth or seventh century. The sophisticated uses of masonry that were introduced during the Roman occupation virtually disappeared during the Middle Ages. In the *Memorials of Ripon*, published in England in the 1600s, it is recorded that one John Mason was paid for making what were called "grund walles" during the period 1399-1400. These ground or foundation walls, built without mortar and quadrangular in form, are perhaps the first documented domestic use of masonry in Britain. At this time masons both quarried and built, developing, or rather localizing, a vernacular building style. Nearly all stonework, up until the introduction of mechanization and contemporary transportation systems, possessed this local individual style.

Early mortar was composed of earth and water, or sand and clay, in either form a bond of dubious quality. The probable reason behind these compositions is discussed in *Romano-British Buildings and Earthworks*:

> In Roman mortar of earth and lime, the latter has sometimes been removed by the action of the moisture of the soil, leaving an earthy residue, which has misled observers into thinking that puddled earth was used for mortar.[2]

Some were not misled, for as early as 1394 there is reference to the use of sand and lime as a coating, both inside and out, on the walls of St. Michael's Church, Bath. In Britain, in Germany and, in fact, wherever rubblestone was employed, some type of finish was applied to the walls.

While earth and water did not a good mortar make, clay and cow dung did, and it was widely used. Exterior coatings of dung were frequently employed on English domestic buildings. It has excellent setting properties, and as an element in coating or pargeting rubblestone flues and chimneys, this humble material may still be recommended. "Pergeny mortar" was the term coined by R. Holme in his *Academey of Armoury* (1688) to describe a kind of flue mortar mixed with dung. This type of mortar was not confined to England; *Mit drecke klieben* is a German saying that translates loosely as, "If there is no chalk, one must daub with dung."

Wattle and daub, although it is a combination of wood and masonry, was an integral part of the masonry tradition of domestic architecture. In prehistoric times, woven basket-work, or wattle-work, had been covered with clay to form a "weathertight" dwelling. The idea was still current in fifteenth-century Essex:

> Long, split saplings of ash and hazel, arranged vertically between each pair of timbers, were sprung into grooves in the upper and lower horizontal beams. They were then tied with green withy bands to short cross-pieces wedged between the uprights. Clay mixed with short straw and chalk was then worked in between the main timbers, thus covering the hurdle-like lathing with a rough "daub" cement.[3]

Daubing was made from various components. It was said that road scrapings were outstanding in their setting qualities. "Clob" or "cob" was basically clay and straw mixed together, and it was pressed into domestic service from at least the thirteenth century. By the nineteenth century in England, the mixing of daub from mud, straw and field stubble was called tempering; at least one 1810 recipe required three parts chalk to one part clay, well kneaded, and mixed with straw.

The terms applied to masons reflect some of these developments in technique. In the England of 1538, those who worked in "cementarry" were daubers, pargeters and rough masons. "Cob mason" was the

Wattle and daub was used throughout the medieval period in Europe. Early settlers brought this method to the shores of North America, but severe climatic conditions in the New Land rendered it ineffective. Plymouth, Massachusetts, reproduction in early seventeenth-century style.

name given to the man who built this sort of structure, while a contemporary of Sir Christopher Wren defined the "white mason" as a hewer of stone and the "red mason" as a hewer of brick.

As more sophisticated domestic forms emerged, the use of local materials continued to govern building styles. The Dutch, Flemish and German skills in building with brick evolved in part due to the high quality clays found throughout these countries. The use of stone and marble continued to flourish in the Mediterranean basin, while the diverse geological makeup of England permitted masons there to become adept in the use of limestone, sandstone, flint and brick.

Building with stone and brick became a necessity in England, because by the seventeenth century the once great forests were nearing depletion. The large amounts of timber needed for squadrons of Elizabethan war and trading ships had taken a heavy toll. Stone was as cheap as wood, if not cheaper as the century progressed. Brick had been used since the fifteenth century as a substitute for stone in non-domestic architecture, and technological advances in manufacturing began to make it readily available and cost-effective. The architect of the aristocracy, Inigo Jones, employed both Flemish and English brick bonding styles in his work during the late 1600s. By 1700, bricks for moldings were being roughly dressed with an axe and then rubbed on a float-stone to attain the desired form. This was the origin of the term "rubbing bricks" to denote fine, well-finished and expensive bricks used on the facade of a building.

Perhaps no other single event made the use of brick and stone more prevalent in England than the 1666 Great Fire of London. The medieval wooden City of London quickly became the brick and stone metropolis of the eighteenth and nineteenth centuries, its skyline dominated by St. Paul's Cathedral and Sir Christopher Wren's fifty-two brick and stone churches. Building with brick was encouraged, both because it was good for the home economy and because brick was less combustible than wood. As the industrial boom of the nineteenth century progressed, London spread, brick by brick, into the surrounding countryside.

The settlement of the North American continent by the French, English, Dutch and Spanish began during the sixteenth century. Ironically, one reason for English exploration and settlement was the need for wood, especially the tall white pine for ship masts. With them these military men and traders brought masons who carried, in turn, the accumulated architectural traditions of the Old Land. As time progressed, European immigrants from other countries arrived, among whom were masons trained in the vernacular building styles of their own lands. Although the Georgian, Neoclassical style, which had reached maturity in England became the dominant style, styles from the rich architectural heritage of other areas also appeared in the wilderness.

Conditions in the New Land dictated the actual appearance of masonry buildings erected in the styles imported from the Old. As a result, a distinctive North American vernacular architecture was born. The use of brick and stone was greatly affected by the skill of the practitioners of the mason's craft and, as it had been in Europe, the availability of raw materials. Both contributed to the shape of the architectural mosaic we prize so highly today.

North American vernacular forms also developed in part because European styles and techniques could not withstand the harshness of an alien climate. The problems were enormous, certainly, but early documentation suggests that building with brick was being practised in the New Land as early as the 1600s:

> The spademen fell to digging, the brickmen burnt their bricks, the company cut down wood, the carpenters fell to squaring, the sawyers to sawing, the soldiers to fortifying, and every man to somewhat.[4]

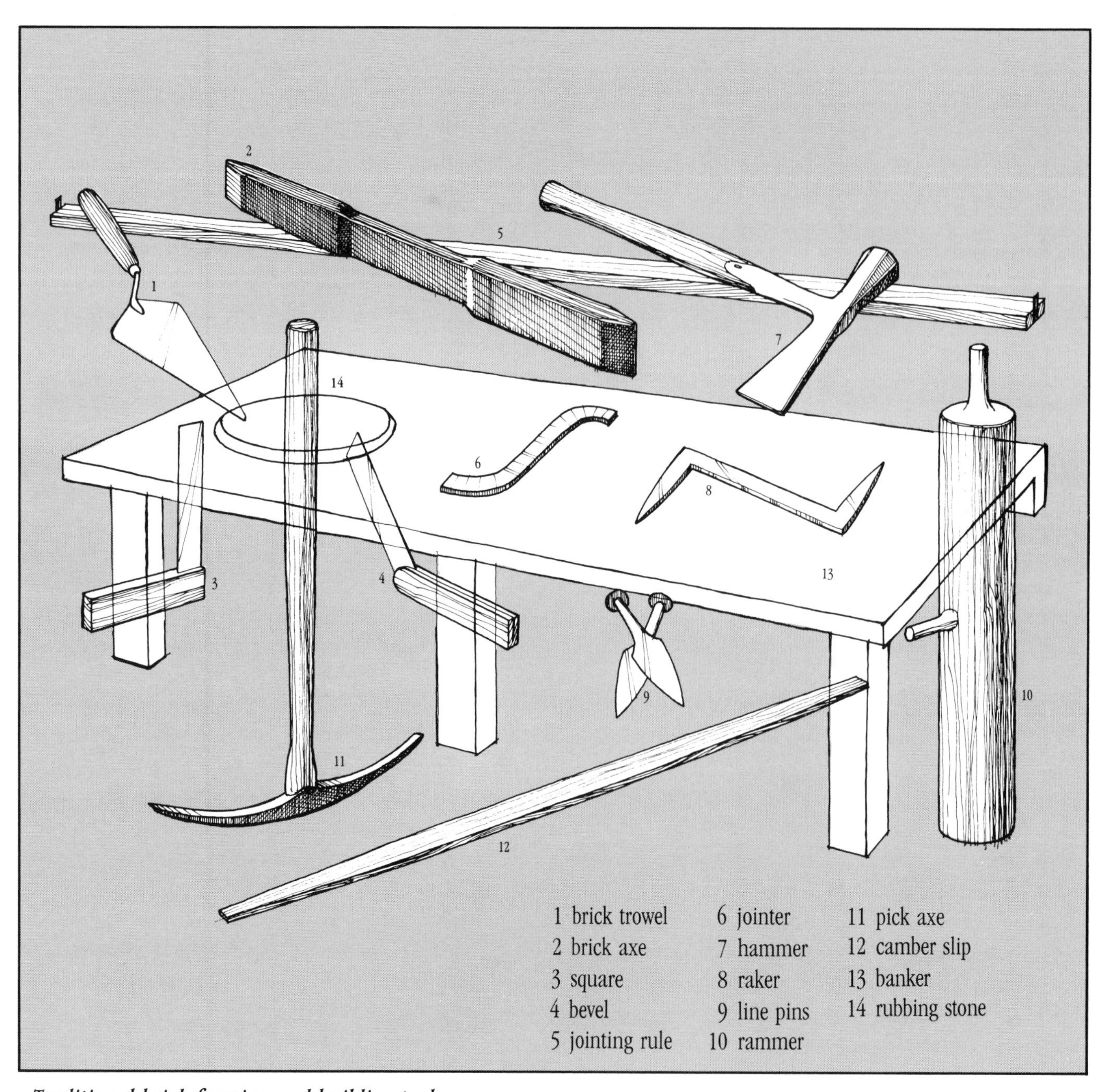

Traditional brick forming and building tools

While these busy residents of Jamestown were building their southern settlement, similar activities were taking place far to the north. In what is now Newfoundland, then called Ferryland, Lord Baltimore resided in a brick "mansion" by 1627. These settlements were peopled largely by the English, but in the southern parts of the continent the influence of Spain was dominant during the seventeenth century. The Spanish colonial style, with its distinctive techniques, adobe bricks, clay tile roofs and lime washes, spans three centuries.

Despite their charm, early brick buildings were often plagued by poor quality materials, especially the lime needed to make mortar. Historian Fiske Kimball cites the unhappy experience of Governor Winthrop, who tried to build a stone house near Boston in 1631: "There came so violent a storm of rain ... (it not being finished for want of lime) two sides of it were washed down to the ground."[5] The problem was understood at the time, and good quality lime was much in demand, witnessed by Edward Scull's 1719 advertisement in Bradford's *American Mercury*:

> Very good lime to be sold by him, next the Franklin Tavern, in Front Street, Philadelphia, at 1/85d per bushel and he will deliver it to any person at Salem, Burlington, or Bristol, at 2s per bushel, if in quantity.[6]

The English were active along the east coast, and the Spanish in the south, but on the northern shores of the Atlantic and inland, the French also erected an architectural legacy. The initial settlements organized by Samuel de Champlain recorded among their company numerous masons and stonecutters. Their stylistic influence would quickly spread from New France as far as present-day Detroit, where in the mid-1700s the presence of good clay for brick and lime for mortar had already been recognized.

French masons and stonecutters had already reached the pinnacle of their craft 200 years earlier during the epoch of the Continental Gothic style. In the New Land they were to leave behind them an architectural record stretching from Louisbourg on Cape Breton Island, up the St. Lawrence River to Detroit, and south through Missouri to New Orleans. The walled fortress-city of Louisbourg was itself an early essay in the mason's craft. Rubblestone structures stood beside timber frame or *collombage* forms. *Collombage* was a framing style imported from France. Like timber frame buildings, this type of structure employed masonry infill or noggin called *pierrotage*, which was made of stone rubble or briquet. Clay for the brick was dug at the Mira River and presumably formed and fired on site. Not all the stone construction was first rate, however, and parts of the walling in the initial construction collapsed. The mason mixing the mortar had used salt water rather than fresh.

During the 1700s, the use of brick and stone spread rapidly in the New Land settlements founded by the French and English. In New England and the southern colonies, the Georgian style was copied and adapted, although in New England wood was used more often than stone because of its ready availability. Most settlers in this area had come from southeast England, a region where stone was not commonly used in domestic construction.

By contrast, the Dutch colonial settlements in New Amsterdam and in the area bordering the Hudson River portrayed the excellence in brick construction for which the Dutch were justifiably renowned. The color and texture of the yellow bricks used in New Amsterdam gave it the air of a typical Dutch town. The use of imported brick and stone, because of their excessive shipping weight, often necessitated finding local sources. Even so, European brick was brought to North America as early as 1661; Holland bricks cost £4, 16s per 1,000, payable in beaver skins. Indigenous industries soon sprang up to compete with imports, and before long the local manufacture

Wall coatings, parging and stucco were used both to enhance the structure and protect it from the weather. Pieces of the stucco on this building have fallen off over the years. Quebec City, Quebec, eighteenth century.

OPPOSITE, ABOVE
The entrance to this Dutch house, with its dual transom light with a nine-pane break-up, four-panel door and sitting stoop, is the focal point of the facade. The Van Alen House, Hudson Valley, New York State, early eighteenth century.

OPPOSITE, BELOW
The angled brickwork in the parapet gable wall of the Van Alen House is known as mouse toothing and is a unique feature of eighteenth-century Dutch brick buildings in the United States.

of brick, roofing tile and fireplace tile was widespread. By 1789 North American brick had become an exportable product. In that year, 129,000 bricks were shipped from New Hampshire to the West Indies.

Although building with brick was probably the earliest method, the use of stone in domestic construction was not far behind. Dutch colonials, the Pennsylvania Dutch (Germans), as well as the French, were early practitioners of this form. At first, French structures were wood with masonry infill, but as such cities as Quebec City grew in size, and segments of the population became more affluent, stone buildings became common. By the end of the 1600s, various large contractors — most had made their fortunes in military construction — were practising in Quebec City. According to a 1739 enumeration of eighty-three houses in the city, fifty-five percent were recorded to be of masonry construction.

In the early 1700s, the process of building a house in New France would involve two parties: the craftsmen who supplied the labor and the client who supplied the materials — stone, lime, interior woodwork and so on. As can be imagined, the already short building season on the Upper St. Lawrence was further shortened by the scheduling of such a complicated operation. To circumvent problems and delays, by the mid-eighteenth century the mason had become the building contractor. Often in partnership with another craftsman, he would employ sub-trades much in the same manner as is done today. A fixed price was quoted, and a finished house was presented.

By the end of the eighteenth century yet another pricing method was employed, what today we would call "cost per square foot." The mason was paid a set price for every *toise* or fathom of masonry wall. Separate prices were worked out for dressed stone, because it was more time-consuming to prepare than the more usual rubblestone.

The construction of a stone house in New France took place according to the following timetable. In the fall, the client and the contractor worked out a

This rubblestone house in Quebec resembles those in Europe in form, material and construction techniques. Rivière du Loup, Quebec, eighteenth century.

design and drew up a contract. During the winter, the masons dressed the stone for window surrounds, door openings, etc. In early spring, the foundation and walls were begun and completed, it was hoped, by late May. By the end of June, the truss system and roof were installed and the entire structure finished, ready for occupancy by November 1. Schedule was dictated by climate, as is still the case in masonry building in the northeastern United States and Canada. The presence of frost renders freshly applied mortar useless, and the wait for spring was as trying then as it is today.

Despite climatic problems, masonry buildings were frequently preferred. The fact that the population of Quebec City, for example, was becoming financially able to afford better housing was fortunate indeed. In 1727 an ordinance was passed in the city forbidding the building of frame houses, even if they were meant to be covered with a lime and sand plaster, because of the monumental fire hazards involved. Deciding on a stone house was not merely one choice among several equal alternatives: stone construction cost six times that of *pièce sur pièce* or timber-frame. While *pièce sur pièce* might cost 300 *livres*, a stone house might cost 2,000 *livres* or more.

Historically, the actual building process, whether stone or brick, would make a mockery of the tidy, elegant "restorations" we so admire today. A foundation and basement would have been dug and the earth spread into the street. Wagonloads of sand and lime would be strewn about the vicinity. Masons continually chipped stone, while others built, and still others mixed troughs of sand and lime. Scaffolding spread out from the tapered walls like a spider's web. French pointing, quite different from the English technique, involved plastering the joints so that only a small portion of stone appeared. More often than not, the entire surface was plastered — a practice treated with non-enthusiasm by today's old house advocates.

The site was alive with activity. Masons, carpenters and joiners were served by day-laborers who mixed and brought mortar, sharpened tools and generally looked after the menial tasks of supplying the master craftsmen. These laborers were not always men. In 1687, two women of "bad character" were to be shipped back to France. The government, instead, sentenced them to hard labor on public works: drawing water, serving masons and sawing wood.

At the dawn of the nineteenth century, the use of stone and brick in domestic construction was well established throughout settled North America. Stylistically, the sixteenth-century works of Andrea Palladio had influenced generations of master builders in Europe. In turn, their influence was widely felt in the New Land during the eighteenth century. To a large extent Palladio himself was inspired by the Roman architect Vitruvius, a gentleman of relatively minor consequence in his own time. Through Palladio, the use of Vitruvian facades, embellished by projecting pediments, pilasters and columns and what came to be called Palladian windows, redeemed and forever glorified his name in the history of English domestic architecture. The transmission of these elements from Rome to such locations as Upper Canada (Ontario) or rural New England may seem unimportant at first glance, but old house enthusiasts should be as aware of the wellsprings of style as they are of the physical details of construction methods.

In North America, military architects and engineers were largely responsible for maintaining these traditions. For example, by the early nineteenth century they were dramatically affecting the evolution of masonry building styles along Upper Canada's Rideau Corridor. Lord Selkirk had already noted the presence of French-Canadian masons for hire in Kingston during the late 1790s, recording their fees:

Stone walls are built about a guinea or 4$ per toise square — twenty inch thick — corners not double measured — mason finding all materials.[7]

But it was not until the War of 1812, and the anticipation of further conflict between the United States and Canada, that the Rideau Canal from Ottawa to Kingston was conceived and constructed. The builders, the Corps of Royal Sappers and Miners, were made up of many nationalities, all "bred to the trades of stone cutters, masons, miners, lime burners, carpenters, smiths, wheelers and gardeners."[8] Cornish masons were renowned for the precision of their stone cutting, while the French, Scottish and Irish were employed for their particular skills in masonry construction. These workers built three barracks of "good rubble masonry" on the site of Canada's present-day Parliament Buildings, the first of many such structures erected along the canal and in the surrounding countryside.

Although the logistical problems in domestic construction were the same for Scottish and French masons, styles and building methods varied. The main

ABOVE
The influence of Andrea Palladio's writings can be seen in the proportions, columns, pediment and sash design in the facade of this building. Cooperstown, New York, last quarter eighteenth century.

OPPOSITE, ABOVE
The ashlar facade of a Neoclassical coursed rubblestone house. Merrickville, Ontario, mid-nineteenth century.

OPPOSITE, BELOW
This uncoursed rubblestone house in Quebec is a good example of the French mason's style. Eighteenth century.

This one-and-a-half story Scottish house epitomizes the use of traditional forms and construction methods. Perth, Ontario, 1820s.

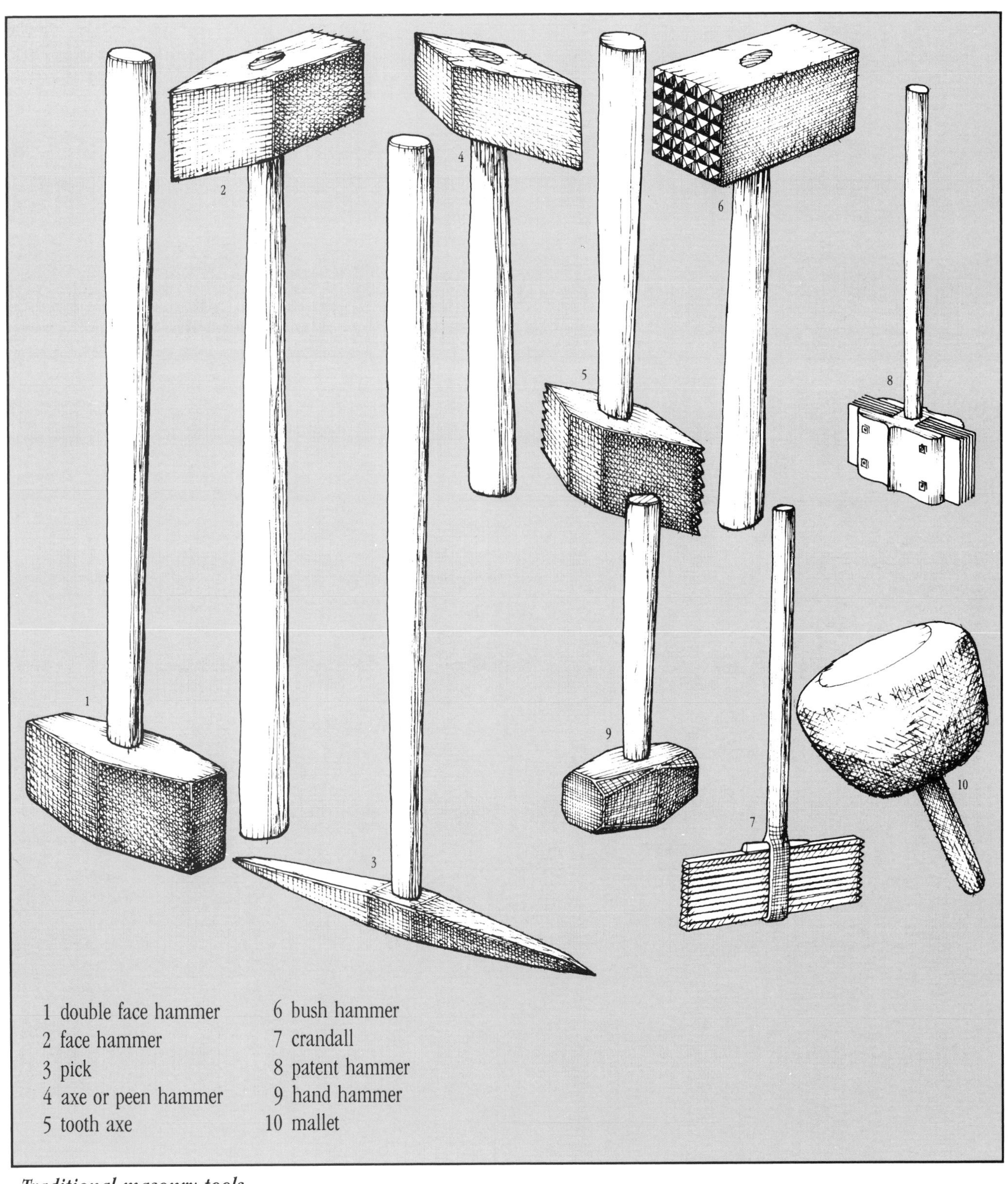

Traditional masonry tools

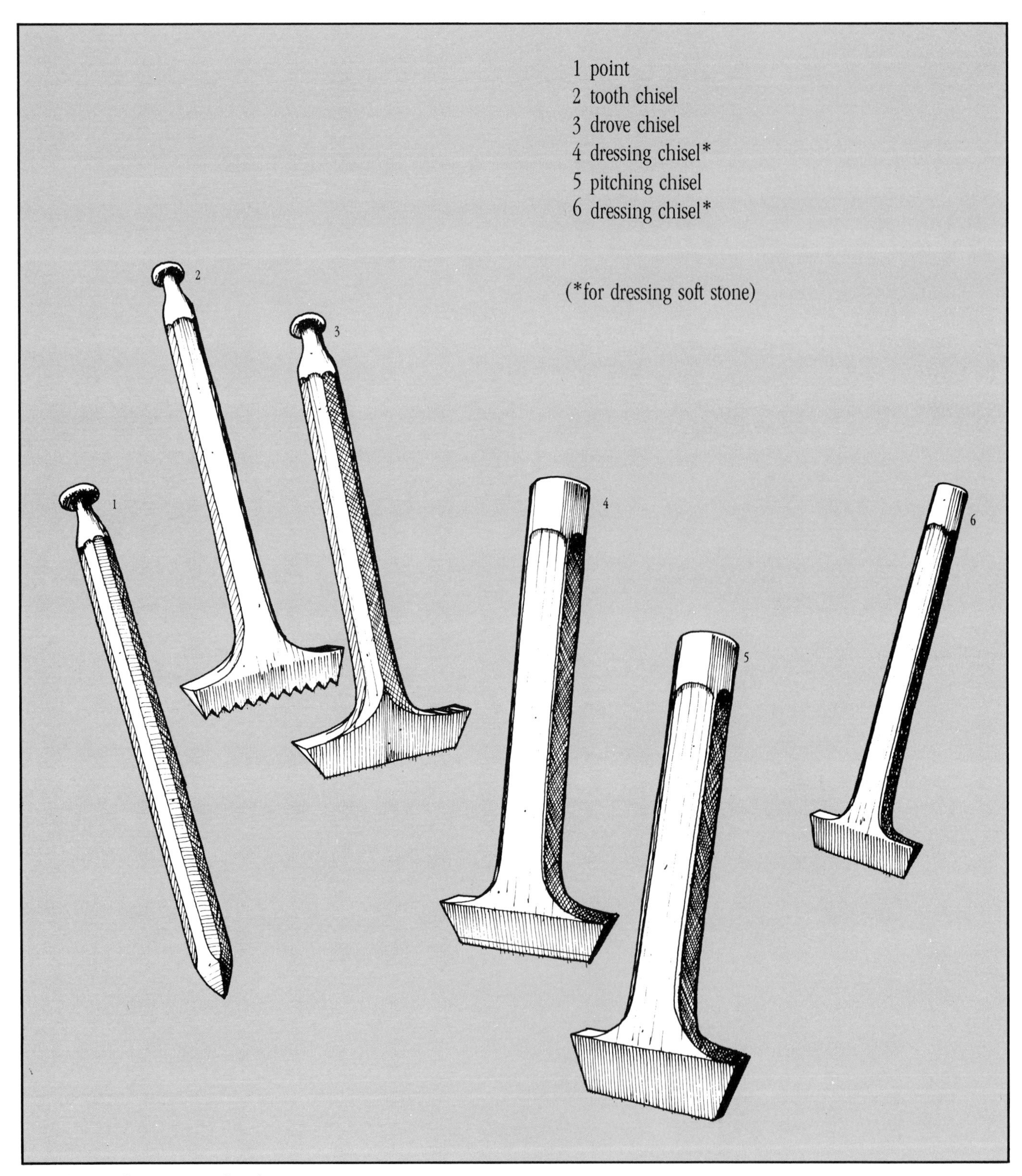

Traditional masonry chisels

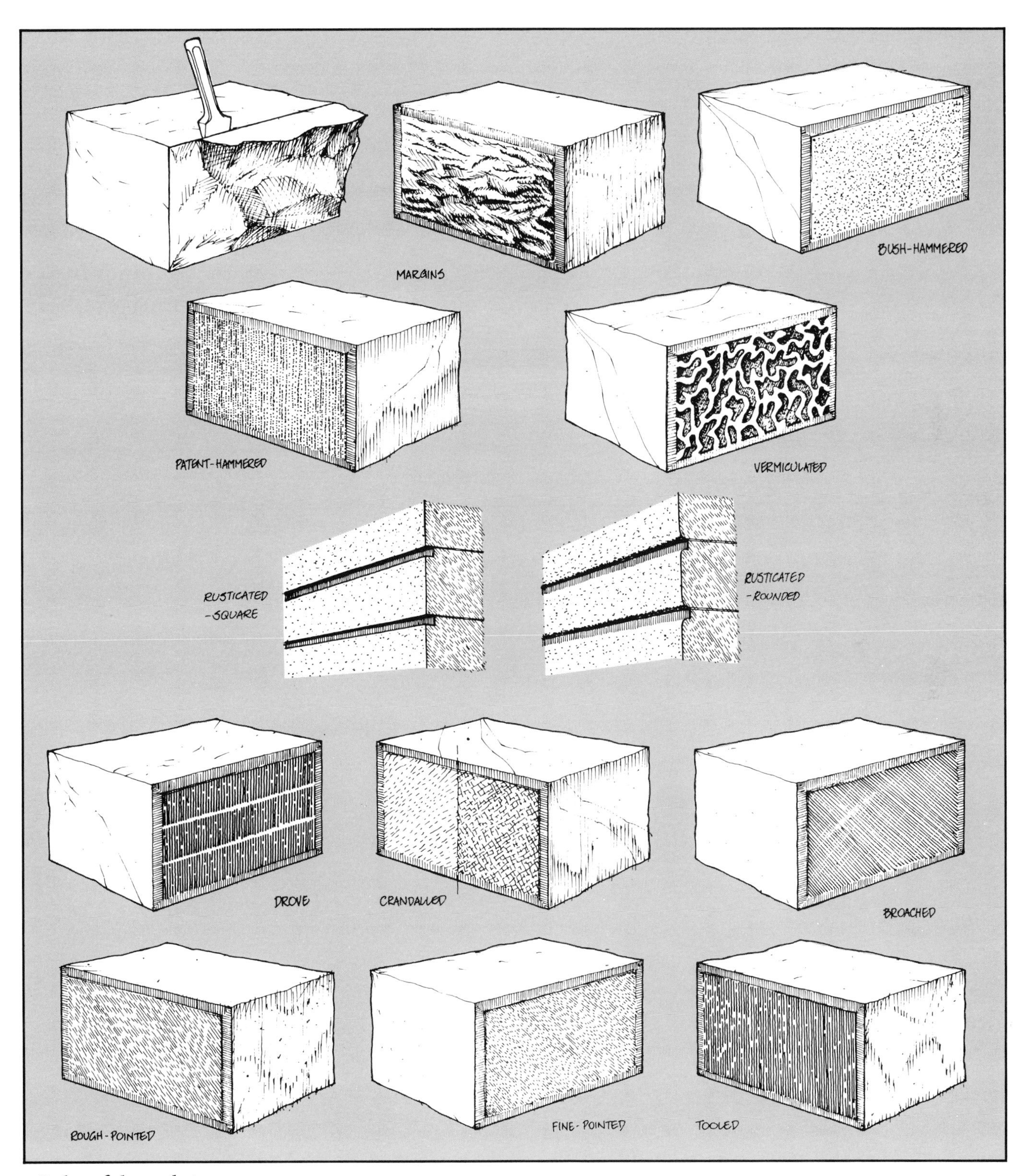

Styles of dressed stone

differences were in wall thickness, roof pitch and jointing. In the late eighteenth century, Scottish settlements in Pictou, Nova Scotia, were the first to mirror their roots. Later settlements in Prince Edward Island reveal the stylistic influence of Edinburgh's Georgian builders. The Irish also had a rich homeland tradition in domestic stone construction; they employed and honed their art throughout these northeastern areas.

The classic style of the period was the one-and-a-half story rubblestone house. If a degree of affluence had been achieved, perhaps ashlar or cut stone would be used in the facade. Quoins, sills and stoops might be dressed and tooled in houses built for the more prosperous.

Dressed stone was valuable and much in demand in the early nineteenth century. For instance, Hector Bradfield's Stone Cutting Shop in Elizabethtown, Brockville, Canada West (Upper Canada until 1840), contained a quantity of 10,000 stones suitable for carving and valued at $1,500. When cut into sheet stone, their value increased to $4,000. Often, the masons' talents were employed on interior fixtures. The 1845 John Greville House in Montague Township, Ontario, has a cut stone and tooled fireplace surround, as well as tooled and styled hearthstones. The sink, inset into one window well, is detailed in a similar manner. Built in the Neoclassical style made popular in eastern Ontario by the Scottish and Irish, the Greville house is a fine example of this form. The methods used in its construction form the basis of the discussion of stone house building in Chapter 4.

Throughout North America the vernacular masonry styles that developed were often dictated by the availability of different masonry units. In the west, the first brick-type houses to be built were made of sod cut into sections two to three feet long and sixteen to eighteen inches wide. Stacked with joints overlapping, these "bricks" provided adequate if not terribly permanent shelters. A style of cobblestone construction was employed in New York State, partially because suitable materials were near at hand. The method was imported from England but was based on Roman practices. Such houses were built as far north as southern Ontario, and a contemporary account gives a clear description of the technique:

> Cobblestones of any size not exceeding six inches in diameter may be used, but for the regular courses on the outside those of two inches in diameter should be preferred. Small stones give the building a much neater aspect. Two inch stones are very neat, though three inch stones will answer. The inside row of stones may be twice as large as those outside .... Mortar ... eight to nine bushels of clean, sharp sand to one bushel of lime stone ... the strength of the building depends on the goodness of the mortar .... The thickness of the wall is sixteen inches, though twelve inches will answer very well for the gable ends above the garret floor.... When the foundation, or cellar wall is levelled and prepared, a layer of two (or two and a half) inches of mortar is spread over it, and the stones are laid down into the mortar in two rows which mark the outside and the inside of the wall leaving about an inch between each adjoining stone in the same row. If the wall is to be grouted (mortar, sufficiently fluid, poured between the stones filling the interstices) the two rows are formed into two ridges by filling the vacancies between the stones with mortar, and the space between these ridges (about a foot in width) is filled with such stones as are not wanted for the regular courses. The grout is then applied. If the wall is not to be grouted, however, the mortar should be carefully pressed around every stone, making the wall solid without flaw or interstice. When one course is levelled, begin another.[9]

Support or out-buildings were frequently built from masonry units. Such structures were constructed to retain either warmth or cold. Ice houses, for example, were used to store ice gathered from lake or river during the winter. The ice was cut, drawn to the house, stacked and covered with sawdust to await the long, sultry months of summer. Milk houses with their thick rubblestone walls and plastered, whitewashed interiors, served a similar purpose. These buildings were often set into the ground, so the floor level was below grade, to take advantage of the insulation offered by the surrounding earth.

RIGHT

A detail of the cobblestone coursing.

BELOW

Although octagonal houses were relatively common, this cobblestone example is unique. Madison, New York, 1840.

LEFT

The milk house was a commonplace domestic structure in the northeastern United States and Canada. This uncoursed rubblestone example was built in the early nineteenth century.

OPPOSITE, ABOVE

Bake ovens were common throughout early North American settlements. This French-Canadian outside clay oven dates from the late nineteenth or early twentieth century.

OPPOSITE, BELOW

A beehive bake oven, so named for its beehive shape. The brick inner form is showing here, because the rubblestone covering has fallen away.

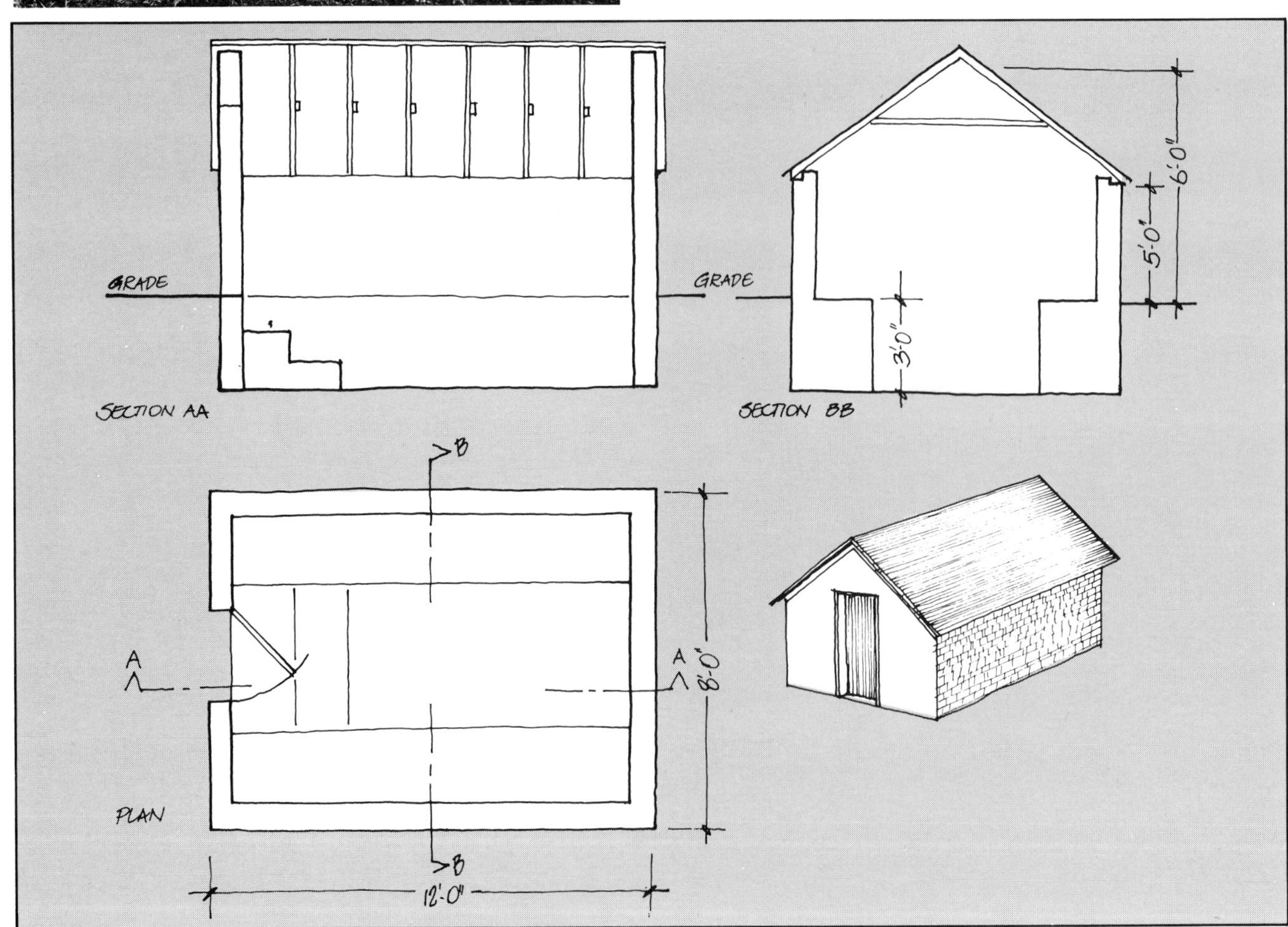

Plan of a milk house

Smoke houses were another means of preserving food; they might be separate structures or, perhaps, attached to but independent of the actual residence. As an alternative during the eighteenth century, an attic smoke oven with an iron door might be built into the chimney stack of the main house. The general practice was to hang the meat from lug poles slightly above the fire so that the smoke would do its job. Another system was to use wooden slats with rows of hooks, called bacon strips, which were inset into the walls just behind the lintel and ran back to the rear wall of the chimney. These strips were common even into the twentieth century.

Perhaps the most nostalgic and frequently discussed masonry cooking structure is the bake oven. It is documented throughout the history of North American settlements. Even in areas where brickwork is seldom found, brick is nearly always part of the oven's lining.

By the mid-nineteenth century, the bake oven had all but been replaced by the cast iron stove, the effects of the Industrial Revolution were widespread throughout North America, and the graceful, symmetrical Neoclassical forms were slowly giving way to the Gothic and Romanesque. Palladio and Vetruvius were becoming less and less important. In their places rose such men as A.J. Downing and J.C. Louden, whose influential building handbooks expounded both current building methods and modern stylistic axioms. Louden, for example, was quite specific in his requirements:

> All the bricks to be used in the building, or brought upon the premises, to be sound and good well burnt grey stocks [bricks made of marley clay]; those to be used in the external parts of the building to be carefully picked of an uniform colour; and the whole laid and flushed solid in mortar; none of the bricks to be slack burnt [imperfectly burnt] or overburnt. The mortar to be composed of the best well burnt grey lime and clean, sharp, pit, or river sand, well tempered together; and to be sifted through a screen.... The walls of the foundations and cellars... to be worked

in brick-work, and grouted with hot lime and sand; the rest of the walls above ground to be of brick-work with a neat, flat, ruled joint.[10]

To some extent Louden's remarks were founded on those of the Englishman Isaac Ware. In 1700 Ware recognized the beauty of brick construction, but he was cautious in his praise:

Many very beautiful pieces in workmanship in red brick [may be found] — this should not tempt the judicious architect to admit them in the front walls of the building.

In the first place, the colour is itself fiery and disagreeable to the eye; it is troublesome to look upon it; and in summer it has an appearance of heat that is very disagreeable.[11]

Ware's aesthetic observations must have had a similar impact on the author of *Village and Farm Architecture*, who recommended,

If for the sake of looks or of preservation, walls of wood or brick are painted (as often they should be) let it still be evident that they are painted brick or wood.[12]

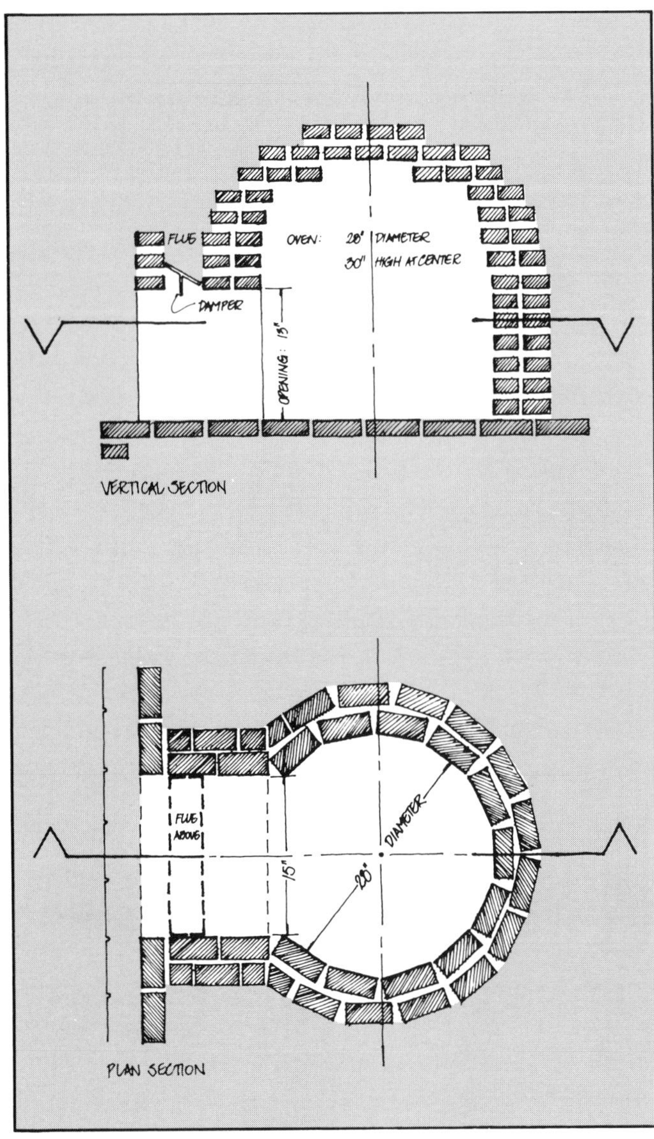

Beehive oven: double course brick construction

Although they are not necessarily in keeping with twentieth-century tastes, these remarks do point out that coated brick or wood was more in keeping with the Victorian era than the natural material.

Industrial developments during the nineteenth century had a significant impact on brick construction. Pre-nineteenth-century bricks were always made by hand and were often quite irregular in shape. As a result, early brick structures frequently could not withstand the North American climate. Champlain, testing the quality of locally made brick in comparison to stone, built a wall "four feet thick, three or four feet high and ten yards long, to see how it would last during the winter."[13] The experiment was probably not a success.

Historically, a brick wall was built using both the stretchers (long faces) and headers (end faces), a style still sometimes seen in our own period. After the footing was laid, the corners were put in place

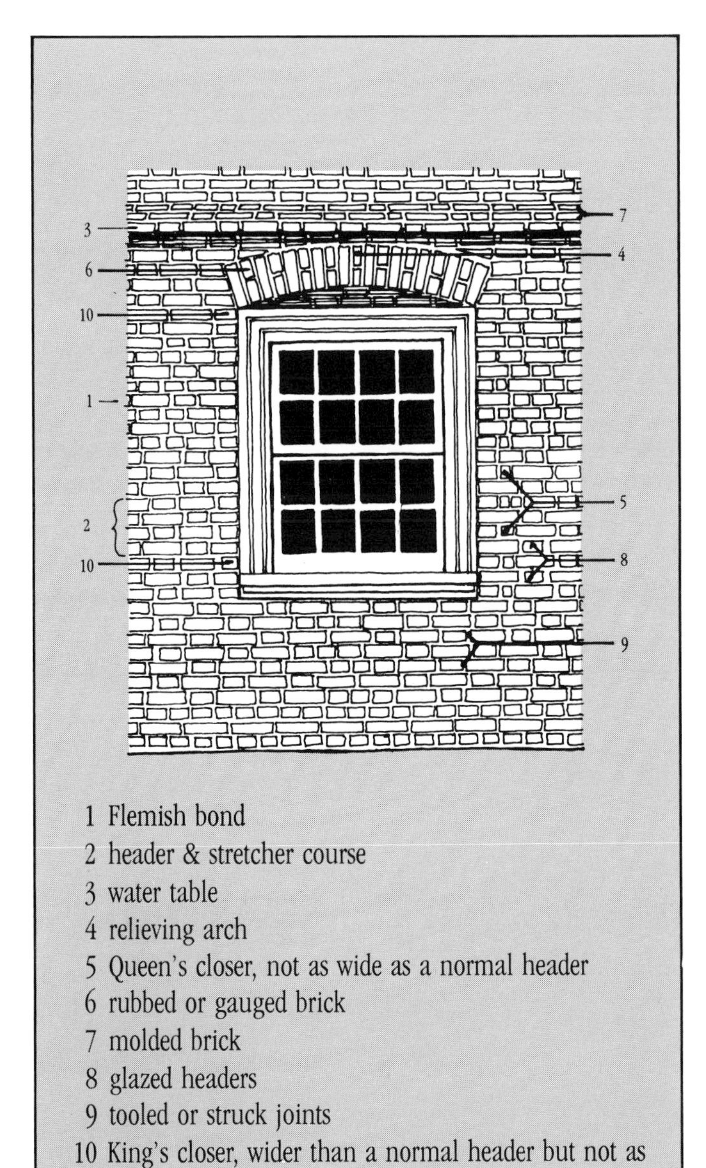

Nineteenth-century brick terminology

with lines set up to keep their measurements true. An outer and inner wall were laid, and the center core filled with bricks, which were shoved or "slushed" into a fairly wet mortar. Because the inner wall was always meant to be plastered or covered in some manner, the inner wall bricks were of a poorer quality than the more finished units used on the outer wall.

The combined walls might vary from nine to twenty inches in thickness, although sixteen inches was the average in domestic construction. Whereas stone varies in shape and requires cutting to produce uniformity, brick is man-made and was always intended to have at least some degree of regularity. Whereas uncut stonework's interlocking irregularities bond it together, the regular system of positioning brick in a wall may be called its bonding.

Bonding is primarily structural in nature, but it has also been used as a visual element throughout the history of brick construction. In the first half of the eighteenth century, English bond was the most common bond in North America. By 1750, the newer English fashion of Flemish bond had been adopted, and it remained in favor until the mid-nineteenth century. Various decorative bonding styles evolved as the century progressed, while some structural variants were developed for specific uses. "Bastard bond," named after the Bastard brothers, was employed in curved wall construction. It was an all header bond, a style that produced a continuous smooth curve with considerable strength.

As the nineteenth century continued, the manufacture of brick became more mechanized and the bricks more regular. However, brick houses were still notoriously damp, as Thomas Jefferson was quick to point out in the 1700s. Dampness was not the only problem: filling the cavity with "slushed in" bricks required considerable labor and considerable quantities of brick. As a result, the hollow or cavity brick wall was developed. (An illustration of this type of wall is on page 36.) It involved two single walls: one

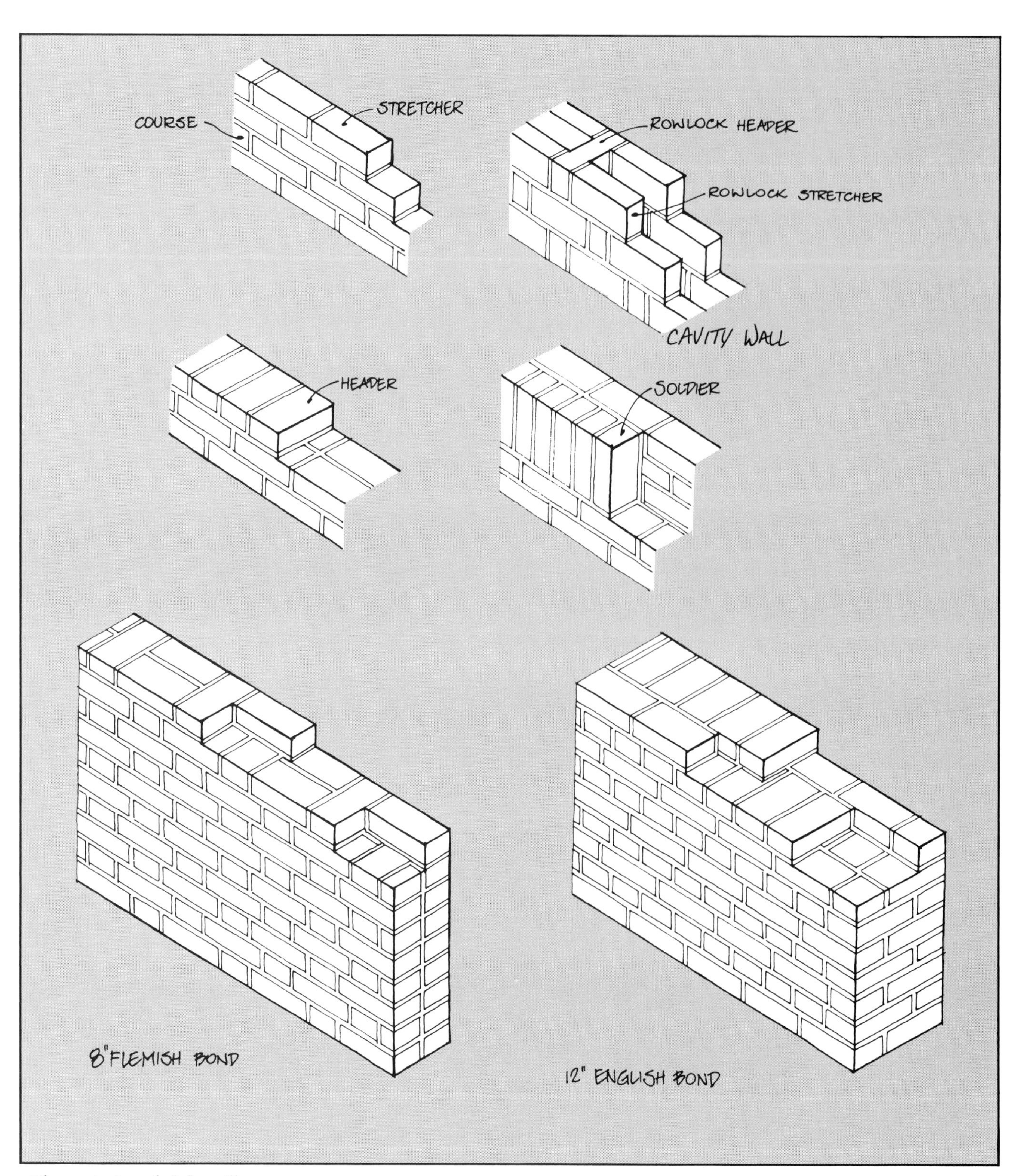

Elements in a brick wall

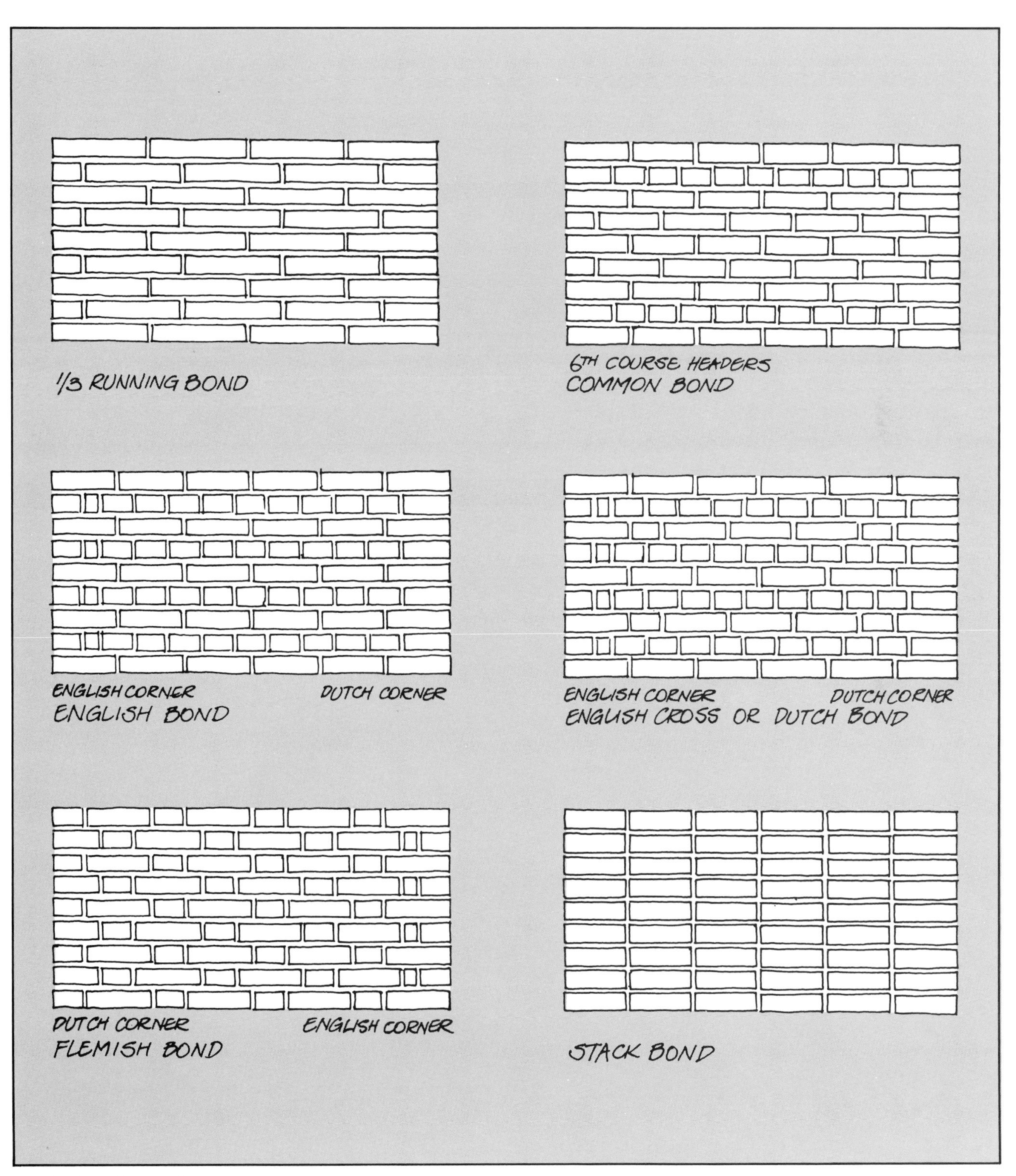

Brick bonding styles

Jack arches

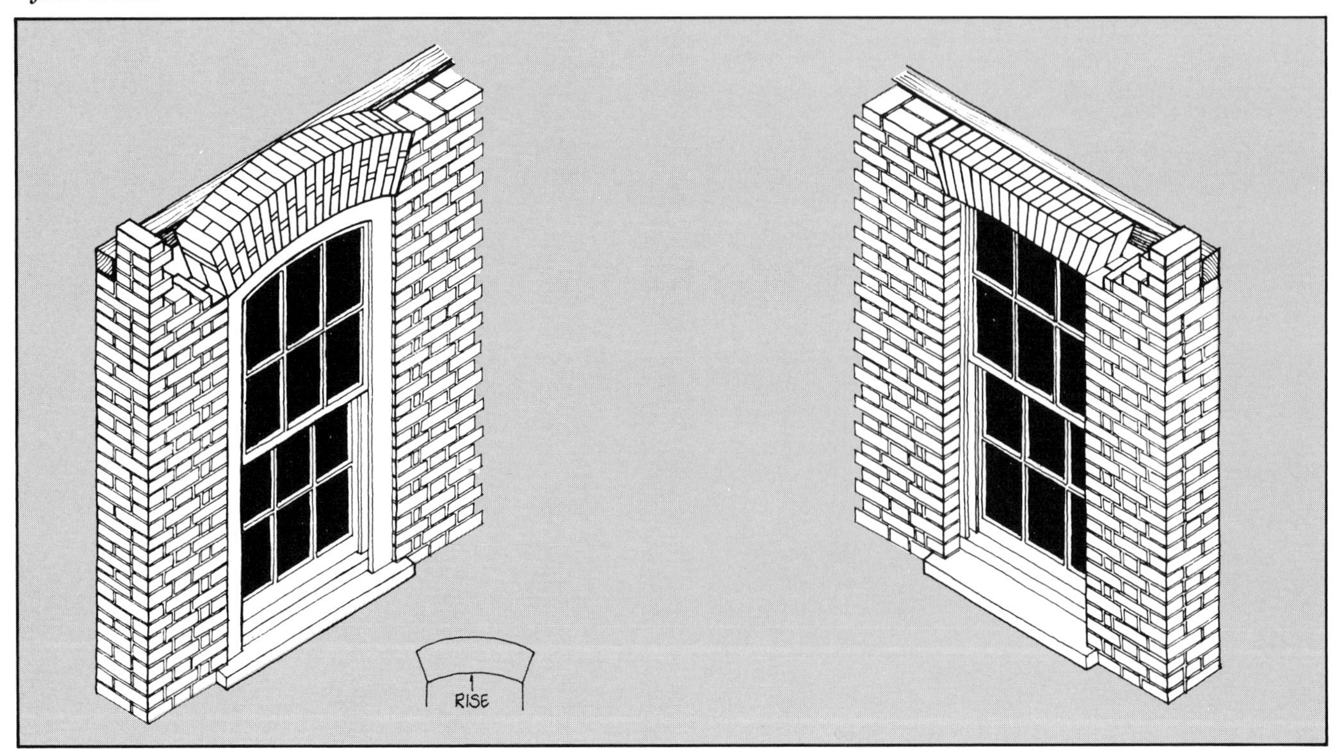

Segmental arches

The Neo-Gothic farmhouse shown in this family photograph indicates that the farm is a prosperous one. The George A. Jebb farmhouse "Fairview," Cookstown, Ontario, circa 1882.

interior and one exterior with a four-inch space tied together with bond bricks or metal ties. A common cavity wall might have been built with two courses on the exterior wall and a single course on the interior. The cavity provided some insulation, and the dampness problem was largely avoided, making it possible to plaster directly onto the interior wall without the need for lath.

The importation of English styles continued into the Victorian era, although there might be a considerable difference in the period in which they were popular in Britain and current in North America. Brick veneer walls, so elegantly perfected in Georgian London, became a mark of North American affluence during Queen Victoria's reign. Today it is not uncommon to find log or timber-frame houses built in the early nineteenth century that have been covered with a layer of brick held on by nothing more than a series of nails to position and hold the brick skin to the wooden framework.

Coatings of various kinds were widely used throughout the history of masonry, both as a preservative and a decorative feature. The primary function of a masonry wall is to keep the weather from penetrating to the interior, at the same time supporting floor and roof systems. Visual impact was secondary to these considerations. Brick in the New Land was not always of the best quality, and practitioners of the bricklaying craft had learned the art of illusion thoroughly from their cousins in England. It was commonplace, for instance, when unable to afford fine, but expensive, rubbing bricks to mix brick dust with mortar and to coat the entire wall so it would be uniform in color. The joints themselves would then

ABOVE

A solid brick wall built about 1840.

BELOW

A mid-nineteenth-century veneer wall showing the nails used to hold the bricks in place.

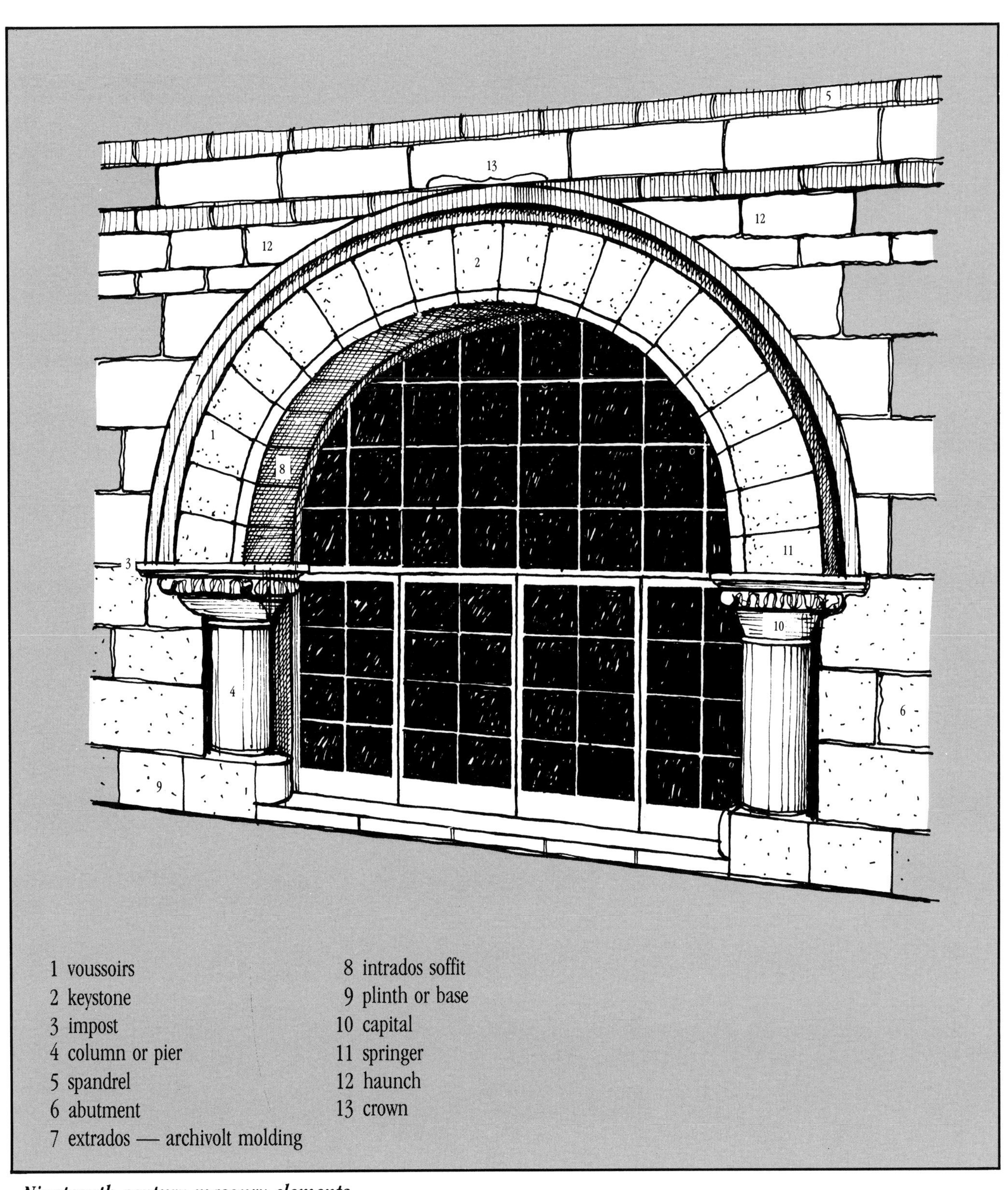

Nineteenth-century masonry elements

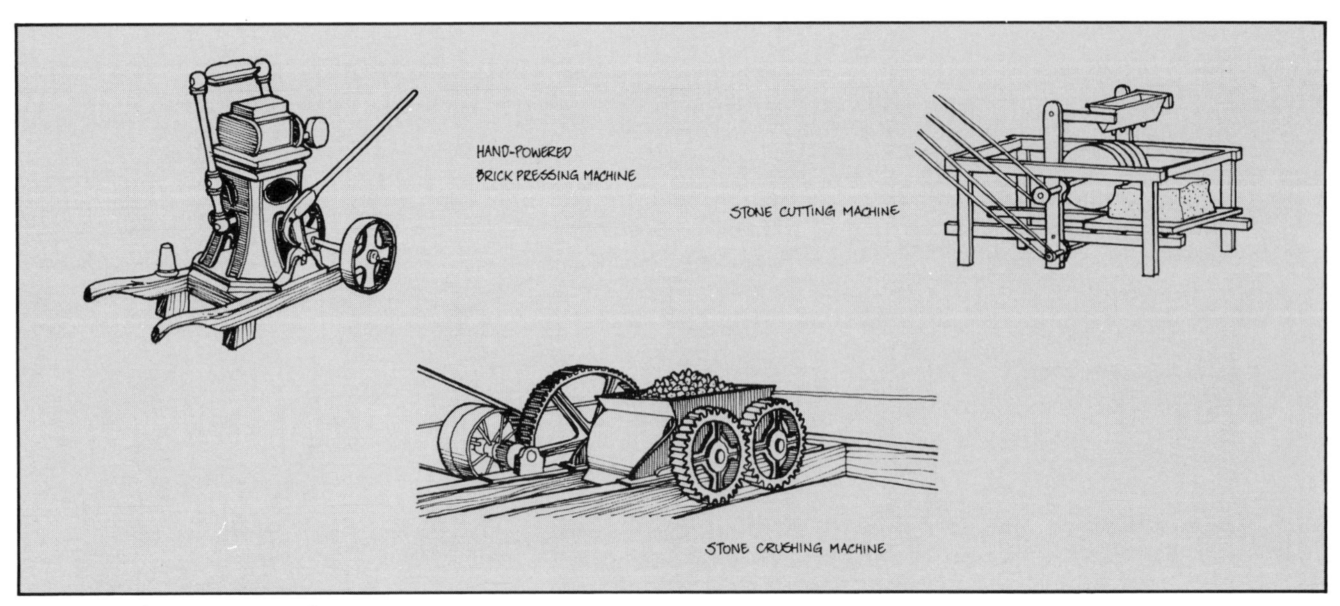

Nineteenth-century machines

be rubbed flush with the face of the bricks, the disguised joint was scored, and imitation pointing was applied using an eighth-inch-deep strip of white lime putty. The method was descriptively called "tuckpointing."

Throughout the last half of the nineteenth century, the post-industrial regularity of brick, the relative ease in transporting it and the stylistic modes of the era made it the dominant force in North American domestic construction. This was the case primarily in urban areas, but its use was also widespread in the country. Cities often had several large brickworks, but small rural manufactories also existed. The Ontario Census of 1871, for example, recorded that a country factory in the Township of Augusta employed eight men, one boy and two horses during a nine-month year, earning $2,700. Its fixed capital was $930, the value of the raw material (clay and sand) was $300, while its 900,000 bricks were worth $5,600.

Building with stone did not entirely cease during the late-nineteenth-century rush to be stylish, but it is true to say in general that the stone buildings surviving today are of an earlier period. Many old house enthusiasts prefer stone houses, both because of their apparent age and because in the best cases they retain the classic symmetry of Georgian architecture.

No matter what the style or the particular construction unit employed, masonry buildings of the eighteenth and nineteenth centuries are fascinating survivors of earlier times. They present very special problems for those wishing to preserve them for themselves and future generations, but the air of permanence, solidity and elegance they possess make the task eminently worthwhile.

This first chapter has looked primarily at the developments of architecture, materials and methods as they appear or were used in the exteriors of masonry buildings. The reasons for this approach are quite straightforward. The mason's craft is most visible here, and if the exterior is not completely understood and made sound in the first place, the entire restoration process is senseless. Preserving a masonry building is usually an attempt to rescue a period environment. If the walls are crumbling, what is the use of replastering the inside? Interior treatments were largely the same for both brick and stone houses. Decoration was dictated by period, but it makes relatively little difference what type of exterior units were employed.

Masonry buildings had a very special role in the fabric of society's past. From the pre-Revolutionary War brick town houses of Philadelphia to the stone farmhouses of rural Ontario, the mason has left us with a lasting monument to his craft. His buildings are a tangible reminder of lives lived during our past. His work, in fact, followed our ancestors from the cradle to the grave as tombstones became an integral part of the masonry industry. The inevitability of death gave his trade an economic stability not often found in the New Land, while the equal inevitability of life gave him the scope to practise his art.

OPPOSITE

The stonecutter's art embraced not only architectural forms but also ornamental forms as this whimsical, naive carving of a tombstone shows. New England, eighteenth century.

CHAPTER TWO

Basic Materials

Building a brick kiln during the first quarter of the twentieth century. Mira River Brick Works, Nova Scotia.

UNDERSTANDING THE MAKING of brick, the cutting of stone, the manufacture of lime and the mixing of mortar is an important tool for the modern preservationist. Important because the keystone in successful preservation is compatibility between what you do and what was done in the past. The appearance of brick or stone buildings in various parts of the North American landscape was dictated by the nature, kind and composition of locally available materials, much as it had been in Europe. This was true only until technological improvements in the nineteenth century made it possible to transport bulky, heavy items successfully and economically.

In a very real way, this localization is a boon to the modern preservationist. Unless you are restoring an elegant, late nineteenth-century town house requiring marble from Tennessee or Italy, the materials needed will probably be readily at hand. Bricks for repairs or the building of a compatible addition to a historic house can often be obtained at nearby brickyards specializing in demolition. One fortunate old house owner was lucky enough to unearth a fair-sized load of brick while digging a sump-pump line in his own yard; finding more of the correct size, shape and color was a relatively simple process. Stone for the same purposes may frequently be found by scouring the immediate vicinity for derelict buildings or barn foundations. The watchword is compatibility.

The process of manufacturing brick, both yesterday and today, may be divided into three basic steps: the digging and working of the clay; the forming or molding of the brick into uniform, rectangular size; and the hardening of the brick by burning or firing. In North America, the most common kinds of clay minerals are kaolinite, montmorillonite, illite and chlorite. Their molecular composition permits them to absorb varying amounts of water. Impurities, such as iron oxide, give the brick its color when fired. In early nineteenth-century London, for example, a chalky clay was used resulting in distinctive yellow bricks called London stock bricks. This classic English brick color, frequently seen in London, was eagerly imitated in North America.

Historically, the clay was prepared for forming by soaking and then working it with a shovel or kneading it with bare feet. Early mechanization took the form of ringpits or pug mills, which would have been fairly common sights in eighteenth- and nineteenth-century North America. After it was thoroughly mixed, the clay would be thrown into molds and the excess scraped off. The bricks were then spread out in a level yard and left there to dry. Once most of the moisture had evaporated, the bricks were fired in a kiln or "clamp."

Firing was a large commercial enterprise or a smaller local venture. It took two weeks to fire a million bricks in the nineteenth-century brickmaking plants on the Hudson River. When bricks were made in less populated areas, scove kilns were used. These, too, were made of brick in a

> rectangular shape with arched openings like small tunnels running lengthwise along the base. Usually it took more than a hundred thousand bricks to make a kiln. Spaces were left between them as they were stacked in order to allow heat to rise freely among them and burn them evenly. The brickmakers piled burnt bricks in an outer cover, or scoving, all around the stack, then plastered them with mud to keep the heat inside the kiln and prevent air from entering. Often they built an open shed over the kiln to protect the bricks from rain. When the mud had dried, a hardwood fire was lit in the arched openings. By careful burning, the kiln tender was able to achieve a temperature of two-thousand degrees F., capable of producing well burnt bricks. Often, however, temperatures varied considerably from place to place inside the kiln, and when it was torn down after the required six days of burning, bricks of varying degrees of hardness and durability were found.[1]

Some of these bricks were so poorly burnt that they had to be discarded. Others had a fair degree of quality in form and strength. These were called common, samel, sandal or clinker bricks and were used for backing, infill or interior walling where they would be covered with plaster. The twentieth-century penchant for exposing common brick interior walls uncovers structural irregularities of little aesthetic appeal. Caution should be exercised when buying used or antique bricks; the common variety is simply not hard enough to be used on the exterior. The best products of the kiln, called stock bricks, were of good quality in uniformity and strength. Also called rubbing or face bricks, they were reserved for use in facades.

Today's brick mason deals with an extremely hard, uniform, standardized building unit. The mason of the eighteenth century, on the other hand, was supplied with materials of some regularity, but he had to shape the softer units to his own needs. A brick axe was employed for rough shaping on a banker or work table, and a rubbing stone was used for finish work. The bricklayer did the actual forming and building, while his laborer mixed the mortar and supplied the necessary building materials.

Part of the mason's job was to use the chosen bonding system in a consistent manner. Bonding styles were developed to give the building both structural and visual integrity; their appearances over the centuries are shown in the illustration on page 37.

The use of brick in domestic architecture was not static during the eighteenth and nineteenth centuries. Crude early bricks may have no more than a rough thumb-gouged line down the centers of their barely rectangular forms, while nineteenth-century bricks from the same kiln may be of surprising uniformity. By the mid-nineteenth century, a masonry material called terra cotta was widely used in domestic architecture. It was similar to brick but finer in texture, made in molds in much the same way as the specialized bricks used for belt or water-table courses. The popularity of terra cotta was assured by the Gothic and Romanesque revivals taking place during this period. The railway reached out across the continent, and decorative tiles, chimney pots and other mass-produced terra cotta components became a common sight.

From manufacture to installation, brick is man-made. Stone, on the other hand, occurs in nature in a great variety of shapes and consistencies. To be useful for domestic construction, the mason must give it at least some degree of uniformity, although in historical terms, at least, stone's irregularities were exploited for bonding strength. Stone falls into three generally accepted major classifications. Igneous rock is formed by volcanic action; a good example is granite. Sedimentary rock occurs when rock material is united with mineral agents and solidified in a process called lithification, as in limestone and sandstone. Metamorphic rock is the result of the recrystallization or realignment of chemicals in a previous rock form, slate or marble for example. All three were used in period North American homes.

In the earliest instances, surface stone was gathered and used for building. As communities became established, a more sophisticated but still fairly simple method of quarrying was employed. Channels were cut at right angles to one another, and wooden wedges were driven into the channels. The wedges were wetted, and when they expanded the stone would split

OVERLEAF

LEFT, ABOVE

Digging clay for bricks using poles called cuckles. Winnipeg, Manitoba, early twentieth century.

LEFT, BELOW

A horse-powered pug mill or clay mixer in use during the early twentieth century.

RIGHT, ABOVE

Laying freshly molded bricks out for drying in Manitoba.

RIGHT, BELOW

Machines such as this one in Saskatchewan mass-produced standard-sized bricks throughout the last half of the nineteenth century.

ABOVE

Stewart's Quarry, Rockland, Ontario, 1898.

OPPOSITE

The late nineteenth-century Neo-Gothic style was ideally suited to the use of mass-produced terra cotta elements.

off from the main mass of rock. This form of quarrying was usually done in fairly close proximity to the surface. Experience led to the discovery that better quality stone could be found some fifteen to twenty feet below grade. Even so, throughout most of North America stone used for rubblestone masonry was quarried close to the surface.

An even more sophisticated quarrying process, still in limited use today, was the plug and feather method. A series of three-quarter-inch holes were made using a jumper drill — a tool similar to a prybar — held by a quarryman who twisted it slightly between blows of the hammer. When the required depth was reached, iron "feathers" were placed in the sides of the holes and a plug was set between them. The feathers prevented the plug from binding, and as the plug was driven true through the line of holes, a block of stone would be cleanly sliced.

Blasting and sawcutting were two other and later methods used in quarrying. In 1846 Augustin Trepannier, a master mason from Quebec, patented a machine for cutting, dressing, sawing and boring stone. His invention heralded a new age in nineteenth-century quarrying. Wire cutting and circular sawing are the techniques used today, but the basic procedures remain the same.

Suitable quarry sites were eagerly sought and their contents recorded, but their exploitation was not always undertaken in a logical manner. One such nineteenth-century quarry in Ontario contained:

12 to 18 inches — purple-banded stone with three inch white top.
10 inches — purple-banded stone.
4 inches — soft, white stone with brown spots.

From this quarry, very large blocks of stone have been obtained — one 30 feet long, 2 feet wide and 18 inches thick is said to have been quarried. Unfortunately most of the good stone has been removed, and, unless prospecting down the hill reveals a further supply, the purple stone is practically exhausted. The practice in quarrying was to break the stone with wedges, very little powder having been employed. With the bedding, the stone breaks freely, and across the beds a good uniform break is obtained by lining and striking with a hammer. Much valuable stone was destroyed and the regular development of a quarry hindered by allowing contractors to quarry their own stone; in consequence of this, the exposure was picked over and no proper quarry was ever opened. Contractors paid the owners $2 per cord for the privilege of operating.[2]

Most of the stone from this site was used in commercial or industrial buildings, although the Code house in Perth, Ontario, is an outstanding domestic example of its quality.

Stone was rough-dressed at the quarry, while finishing was usually done in the masons' "shops" during the winter months. As in Europe, master masons in North America employed and guided free masons who carved and rough masons who shaped the quarrier's blocks. In 1695, for example, Francois De Lajoue signed an agreement with four stonemasons to dress stone throughout the winter, and he agreed to pay them:

12 *sols* for each foot of cut stone with moulding (*en Capucine*).
20 *sols* for each *pied* of dressed stone, intended for windows, doors and fireplaces.
55 *sols* for each workday employed in laying masonry, half payable in goods and the other half in money.[3]

Rubblestone construction routinely required minor dressing while the building was in progress. Dressing and tooling stone are fascinating and exacting procedures. The products of one stoneyard are illustrated on the next page.

No matter which type of masonry units were employed, whether brick or stone, some type of bonding agent was ordinarily used to hold them together in a coherent mass. Oddly enough, as late as the 1840s the sand-clay mixture used in fifteenth-century England as a poor copy of Roman mortar appeared as the bonding agent in North American mud brick buildings.[4] This is not to say that lime in mortar was unknown. The earliest record of lime

The stoneyard of George Oakley, Toronto, Ontario, 1887.

manufacture in the New Land comes from seventeenth-century Providence, Rhode Island. In 1662, a Mr. Hacklet "applied to the town for liberty to burn lime, and to take stone and wood from the commons for that purpose."[5] Mr. Hacklet probably obtained the lime by combining limestone and wood in a large heap, burning it and then sifting as many ashes as possible from the mixture.

A more detailed explanation of the process is contained in a description of seventeenth-century Quebec where woodcutters were also the lime burners:

The lime used for making mortar, roughcasting and plaster was produced by heating crushed limestone in a conical oven. With the moisture and carbon dioxide driven off, the material became quicklime which was slaked with water after being removed from the oven. The hydrated lime was being measured by the hogshead (*barrique* or pipe), and it seems reasonable to believe that it was packed into casks for delivery to the work site. Lime left exposed to the air would absorb carbon dioxide and revert to limestone which was useless to the mason.[6]

Limestone was not the only source; in maritime regions oyster shells were commonly substituted. What was needed was calcium carbonate. Burning lime with some degree of sophistication, no matter what the raw material, benefited from the mechanization

The manufacture of lime in kilns like this one was a major industry throughout North America before Portland cement came into widespread use in the early twentieth century.

taking place during the late eighteenth and early nineteenth centuries. J.C. Louden, for example, spent considerable time in his *An Encyclopedia of Cottage, Farm and Village Architecture and Furniture* describing the construction of a lime kiln:

> The walls should either be built of fire-brick, or firestone; but they are sometimes built of limestone of the same quality as that to be burned within; but having the stones in large masses, so to prevent their being as much affected by the heat as the smaller stones in the kiln, which are mixed with fuel. The upper part of the kiln may either be arched over, or covered with cast-iron joists and flagstones; leaving square or longitudinal holes for admission of air, which may be covered with a plate of cast-iron, regulated so as to give the exact degree of draught required. This contrivance will be found cheaper than the conical kilns of Booker; and, where there is a kiln-shed, it will answer equally well. When there is not a kiln-shed, Booker's covers are undoubtedly preferable; as they not only keep in the heat, but throw off the rain. The shed over the mouth of the kiln is of the greatest use in keeping dry the stones and fuel before they are thrown into the kiln; and not only keeping them dry, but heating them, and thus evaporating part of their moisture.[7]

Until the late nineteenth century in North America, the use of lime-based mortars was predominant. Earlier, in 1824, the Englishman Joseph Aspdin had taken out a patent on Portland cement. This mixture, low in absorbency and harder when set than traditional lime mortars, quickly became very popular among English masons. Technology was capable of producing a harder brick by the latter portion of the century. The two elements were compatible and highly recommended for use in domestic and other buildings.

In dealing with historic buildings of the pre-cement era, it is important to remember that mortar is softer than the surrounding masonry units. When redressing mortar-related problems, mixes of similar quality must be employed both for visual accuracy and to retain the elasticity inherent in mortar. Masonry buildings move; hard mortars such as Portland cement will cause the relatively softer masonry units to fragment.

However, Portland cement is strong and good for fast building, if used as part of a larger mixture employing lime. Masonry cement, on the other hand, is a contemporary, ready-made material that is reasonably effective but prohibitive to use because of cost and the lack of any historical context. Traditional lime-based mortars, as described in Chapter 9, are used by serious preservationists because of their workability, water retention and plasticity.

A constant of all mortars, traditional or contemporary, is the presence of lime. The lime we use today has been hydrated and comes in a dry powder. Historically, lime was hydrated by soaking or wetting, thus making it into a paste with a reputed very high degree of plasticity. Some masons and preservationists still buy calcium oxide and slake it themselves (see Chapter 9 for methods).

Hydrating lime, dressing stone or making brick may be far beyond your willingness or abilities, but understanding these processes is vital when approaching the historic building in need of preservation. Repointing a stone wall with Portland cement, for example, is not only historically incorrect and unattractive, it can actually do permanent damage to the structure. Replacing crumbling face bricks with the common variety will only cause future problems. Besides, if you are anything like me, that single rough common brick in an otherwise flawless period wall will draw your eye and a sigh of dissatisfaction every time.

CHAPTER THREE

Inspection and Maintenance

"Keeping up with the Joneses." A stucco coating tooled to resemble ashlar in a Neoclassical form covers a basic log house. Quebec, first quarter nineteenth century.

WHETHER YOU ALREADY OWN a period house of brick or stone, or are considering buying one, the inspection and maintenance procedures outlined in this chapter should be a central part of your evaluation of the building. Once the house is acquired, an annual inspection of the property is mandatory. Like most inorganic, man-made things, stone and brick houses will not repair themselves. Small problems quickly become larger ones, and the expense grows in proportion.

The exterior of a building is designed to fend off the weather, in fact to repel moisture of any kind. The intrusion of moisture into the masonry building is the chief problem that will confront you. From the roof to the basement, you should ensure that the structure is water-tight.

The roof is the obvious place to begin your inspection. Cracked chimney pots not only expose the chimney to moisture penetration, but if the wind blows them off, they also are potentially hazardous to the roof and to any unfortunate soul walking below. Chimney capping stones should be well fixed to the main stack and the chimney itself well pointed and sound. Poorly maintained flashing, both around the chimney and in other areas, must be caulked or replaced.

..Examine the eaves for possible leaks. Winter ice buildups lead to lifting shingles and moisture penetration. All fenestration (windows) and doors should be checked for adequate jointing and caulking. In areas where the charms of winter require the use of salt as a de-icer, check around the foundation up to about waist level for the presence of efflorescence. This phenomenon appears as salt moves through the inherent moisture in the masonry and is deposited by capillary action on the building surface where it appears as a white stain. Although often not a major problem in itself, the constant use of salt over time may lead to mortar and masonry failure.

The house that is over one hundred years old frequently suffers from a buildup of earth around the foundation, changing original drainage patterns. One 1860 house situated close to a road built in about 1930 gains nearly an eighth of an inch each year from sand used on the road in winter. Water runoff from the house and the surrounding landscape often makes its way into the basement or crawl space turning it into a pond. Rubblestone or brick foundations, with their lime-based mortars, frequently become no more than a pile of building units as the water leaches the lime from the sand leaving only a soft powder. Overgrown basement windows or exterior entrances must be examined for similar water-related problems.

As the interest in and popularity of old houses grow, the techniques employed in rejuvenation may expose the structure to new problems. The major causes of concern that should be added to our checklist are incorrect pointing using improper mortar and the indiscriminate application of modern cleaning methods such as sandblasting. The perils of these "restoration" techniques are covered in later portions of this chapter.

OPPOSITE, ABOVE LEFT
The truss system and rafters formed an integral part of the masonry wall in Quebec houses of the seventeenth and eighteenth centuries.

OPPOSITE, ABOVE RIGHT
Parapet walls were a response to the need for fire protection rather than statements of architectural style. New metal flashing is being installed here in conjunction with a new, period cross-hatch metal roof. Quebec, late eighteenth century.

OPPOSITE, BELOW LEFT
This early nineteenth-century rubblestone chimney displays potential and actual areas of deterioration in both the worn-out pointing and flashings.

OPPOSITE, BELOW RIGHT
Inadequate maintenance of gutters has resulted in major rotting in the soffit and fascia of this mid-nineteenth-century coursed rubblestone home.

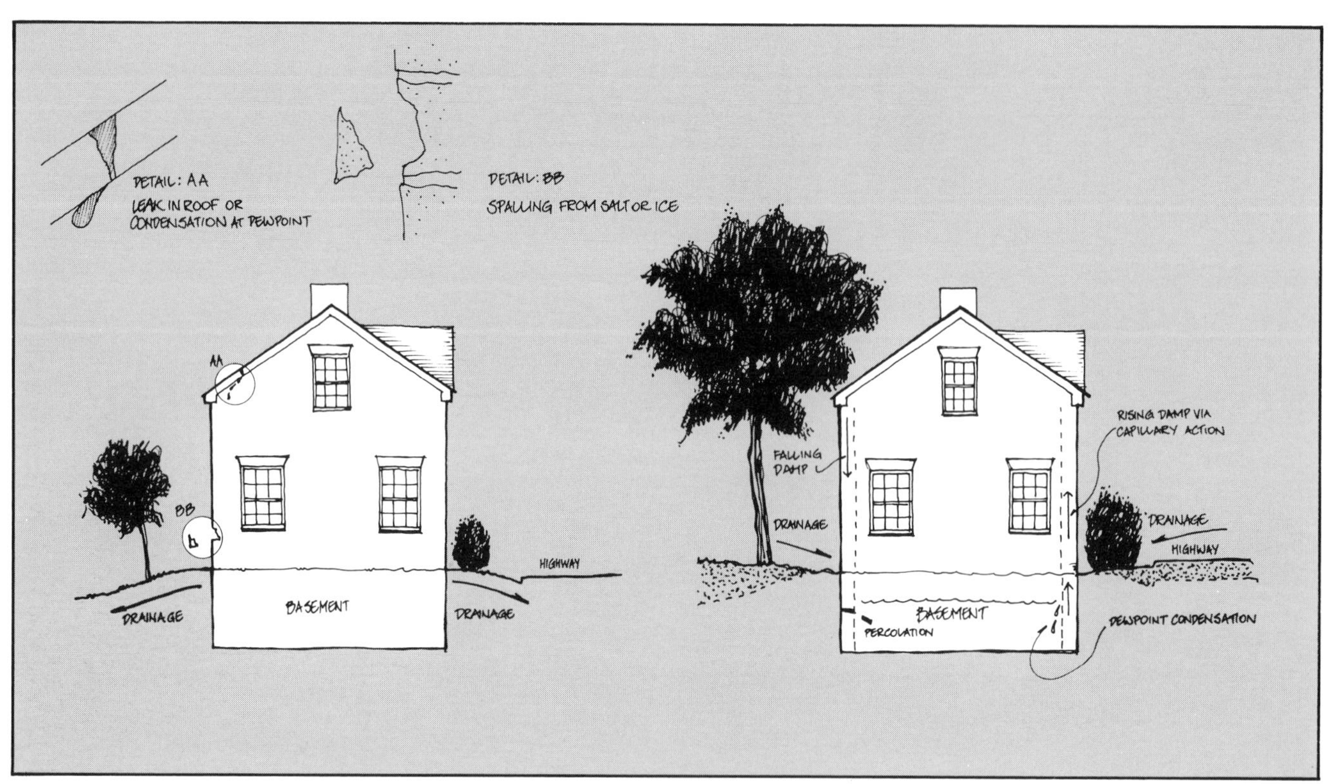

Landscape changes over the years eventually alter drainage patterns

A thorough examination of the walls of a masonry house can be broken into three classifications: structure in general, joints in particular, and the stones or bricks themselves. General structural problems in masonry walls are often very obvious, very difficult to remedy, and therefore very expensive. After the initial construction, some settlement will have occurred as the new structure seated itself in the ground. Theoretically, this will have happened in a uniform manner, but variations in load acceptance are common. Over the years, changes in lifestyle within the house or changes in patterns outside may lead to serious structural problems. Excavations for roads, sewers or even new housing can upset the original footing and foundation. Moisture from storm drains or a built-up landscape may penetrate the foundation, weaken it and prevent it from carrying its load. Mod-

OPPOSITE, ABOVE LEFT
Efflorescence.

OPPOSITE, ABOVE RIGHT
When groundwater flows down through this overgrown entrance, it will enter the basement and cause a multitude of problems. Note the worn-out pointing.

OPPOSITE, BELOW LEFT
Heat loss around the window has caused condensation (moisture) which has rotted the interior wood lintel. As a result, the arch has dropped, cracking the brickwork.

OPPOSITE, BELOW RIGHT
Ground buildup has all but covered this basement window opening.

Condensation occurs when warm air hits cold air. The moisture that is created freezes and thaws, resulting in major spalling as in the brick lintel shown here.

ern traffic may cause vibration a thousand times greater than the original horses and wagons, leading to severe cracking in foundation walls. The area surrounding the historic house was often planted with trees and shrubs in a relatively immature state. In fifty years these plantings have matured and left the surrounding earth devoid of moisture. The result is collapse of the ground around the foundation.

These problems can usually be readily identified, but one area of wall deterioration is not so obvious. Wood lintels were frequently employed in conjunction with stone or brick. Moisture penetration may have resulted in severe rot, rendering these elements useless. All the weight has been transferred to the masonry units surrounding them. This may cause spalling, where the face of the stone or brick has literally fallen off in a sheet; sheer cracking through the depth of the masonry unit; or, if the stone above the lintel is a solid unit, it and the sill below may crack as well.

Heavy timber floor joists can rot in a similar manner. In a badly deteriorated stone house, it is often the case that interior framing and floors are supporting the joists, rather than the opposite. Negligent "renovators" before you or nonchalant tradesmen may have ruined structural support systems by removing load-bearing walls and randomly butchering joists. Wooden lintels and joists may be checked for rot by probing with a penknife.

In the basic outline that follows, structural damage can be ascertained rather easily. These procedures are particularly useful for those who are thinking of buying a masonry house but want to have some idea of the amount of work involved. Stand at each corner of the house and look straight down the plane of both the front and side wall. They should be without bulges or indentations. Measure from corner to corner and end to end of each wall. If there is a discrepancy, some bulging may be occurring that you cannot see easily, extending the measurement in one direction or another. Using nails and mason's string, place strings across the facade of the building both horizontally and vertically. They will clearly indicate any variations in the flat plane of the wall.

Obvious to the naked eye will be wavy joints in brickwork and cracking in both stone and brick. Inside, recent wallpaperings will show telltale tears if the house has structural problems. Older cracks, whether inside or outside, are usually impregnated with dirt; new cracks are bright and clean. To check if a suspicious crack is still expanding, mix up some plaster and fill the crack at its narrowest point. When dry, inspect the patch to see that no cracking has occurred during the drying process. This will tell you nothing. Keeping track of the date, return in a week to ten days. The results should be self-explanatory.

As a final test for stone walls, strike them with a two-pound hammer in areas of bulging. If it strikes a hollow note, there is a void and a problem. If it bounces back with a thud, assume that the wall is sound. If the wall collapses, move.

After you have looked at the walls in general, inspect the joints. Masonry joints suffer as we all do from the ravages of old age; quite simply they wear out. Failure leads to water penetration. If the mortar is soft and crumbling, or missing altogether, you will have to repoint the entire building as described on pages 64 to 73. Areas that have been pointed using hard or inappropriate materials should be cleaned out and repaired. They will be fairly easy to spot, because the consistency and color of the work will be considerably different than the surviving original pointing. The pointing of a masonry building is one of its chief weatherproofing systems. It is safe to say that with adequate maintenance of the joints, using traditional tools and materials, the brick or stone house should last forever.

In addition to the problems associated with joint failure, the bricks or stones that make up a building can suffer from physical, chemical and organic attack. Over the years, abrasive elements carried by the wind (dust, dirt, sand) act as a powerful sandblaster in slow motion. A gradual wearing-down process takes place in the face of this imperceptible physical assault. Although in itself not a major cause of deterioration, in conjunction with joint failure, it can form avenues for moisture penetration. Before 1900, unburnt bricks contained approximately thirty-five percent of their own weight in water. When burnt the ratio decreased to twenty to twenty-five percent. These earlier bricks were soft, porous and much more susceptible to physical attack than most contemporary brick or stone facades.

The most grievous injury sustained by both brick and stone buildings in the northeastern United States and Canada is caused by the freeze-thaw cycle. Poorly maintained joints, worn-out masonry units, inadequate vapor barriers causing condensation, in fact anything leading to moisture penetration are all supporting actors in this drama of the fall and spring. Moisture enters a wall, freezes, expands and shatters the stone and brick. The water melts, freezes again and the process continues. It is not difficult to imagine the end result.

The ravages of chemical attack may well cause the owner of a period masonry home to take up the banner of environmentalism. Though the effects of chemical agents in the atmosphere are visible on an individual house, the problem cannot be solved individually, only en masse by changing our priorities. Until the advent of the Industrial Revolution, air pollution was rarely severe enough to affect masonry materials. As the machine and industry became universal, vast amounts of gas were released into the atmosphere. The Sphinx has deteriorated more in the past twenty years than it had in the previous two hundred.

The main villain is sulfur. It combines with oxygen to form sulfur dioxide, which in turn combines with water to form sulfuric acid. Precipitation that contains this acid is commonly called acid rain. It is no wonder that the famous fogs of London were described as yellow by nineteenth-century observers. Fed by hundreds of thousands of coal fires, they doubtless contained a high concentration of sulfur.

Winter snow remedies are not the only sources of salt; it is contained in everything from bird droppings to the actual composition of Portland cement and brick. Moisture carries these soluble salts into brick or stone where they form crystals. Because they take up more space when crystallized, their expansion can damage masonry units. The most common result is spalling in brick; the face disengages exposing the porous interior, where white efflorescence is often visible. To make certain you are dealing with salt, dip your finger in and taste it.

The third destroyer is organic attack, deterioration caused by plant life. In itself, the presence of plant organisms is not particularly harmful, except for the unsightliness of and minor moisture retention in mosses and lichens. More serious problems arise from vines, roots and branches wearing away particles of mortar and splitting joints. A vine-covered brick wall may look charming, but over time moisture, the primary element in deterioration, will invade.

There are other sources of deterioration not directly due to either age or attacks of various kinds. Occasionally, one is met with a house of historical interest, pleasing style, delightful location but built using appalling construction techniques. The stone may have developed cracks or blemishes during quarrying or transportation to the site; deterioration would have begun at the outset. Bricks may be of poor quality. Even in the finest houses, the facade would be built using top quality bricks, while the side and rear walls used progressively worse materials and construction techniques. Individual stones may not have been laid as they were cut in the quarry, with the stones' bedding plane horizontal. Some New York brownstones peel in layers because of this. Edge-bedded stones were laid with the bedding planes vertical and perpendicular to the face of the wall, leading to wash-out between the layers. This particular problem is more noticeable in cornices and string courses where erosion would act more rapidly than if the stone had been laid on its natural bed.

Repointing

The primary reason for the inspection procedures in the first part of this chapter was to determine the source, amount and cause of moisture penetration. Pre-1900 masonry buildings in North America were constructed with a soft, lime-based mortar. This is the element most susceptible to deterioration due to age and lack of maintenance. Even if your newly acquired period masonry house is sound in all other respects, the chances are very strong that the mortar joints between the masonry units will have to be repaired. The process is called "repointing" or "tuckpointing."

It is important to keep the weather in mind when contemplating a repointing job. Traditionally, the North American mason's year began in April and ended in December, because it is best to do masonry work between forty degrees Fahrenheit (4°C) and eighty degrees Fahrenheit (27°C). Colder weather brings frost, which renders mortar useless. Warmer weather causes the rapid evaporation of the water in mortar, resulting in ineffective bonding. A simple method of ensuring dampness during hot weather is to tie undyed burlap bags together into a large mat, tack it to a spare stud and hang it over the wall on which you are working. The mat should be hosed down at regular intervals, but don't overdo it. Excessive moisture will cause new mortar to run, staining surrounding materials.

Both repointing and tuckpointing use the same procedures. Loose or soft mortar is cut back approximately two-and-a-half times the width of the joint until a firm area is found. A new mortar compatible in material, quality, color and texture is installed and tooled in the appropriate period manner. The process is of twofold importance. Primarily, it weatherproofs the spaces between stones or bricks. Secondly, through details of color and tooling, it visually portrays the texture and harmony of the original builder's vision of the house. Repointing, though labor intensive, may be undertaken by the novice preservationist without fear of technical inadequacies, providing there is careful planning

The example discussed here is a one-and-a-half story rubblestone house with two uncoursed rubblestone chimneys and segmented stone lintels over doors and windows. The basic pointing procedures outlined here are the same for brick, although the tools differ somewhat. These differences are discussed in the appropriate places.

The approximate dimensions of the house are thirty-two feet long by twenty-two feet wide. The tools for the job include the following: a two-pound hammer, a mason's chipping hammer, a cold chisel, a pointing trowel of the appropriate width for the joints, a mortarboard, unprinted and undyed burlap sacking, three or four rubber pails of the farmer's or mason's variety that are marked with measurements and won't break if they fall when full of mortar. You will also need a mixing trough or mortar mixer, a wheelbarrow, a shovel, mason's hoe, hose, scaffolding, safety equipment (hard hat, goggles, work boots, work gloves). The need for safety equipment is obvious; lime-based mortar will eat holes in your hands unless they are conditioned to manual labor. If any of these tools are unfamiliar to you, you may wish to consult the illustrations on the next page.

Once the tools are in order, the next step is to assemble the proper materials: dry, fine masonry sand, preferably taken from the same pit as the original, but at least having the same quality and texture as the original; mason's hydrated lime; and white Portland cement. White Portland is not the same as ordinary Portland, and the average lumberyard may have to order the lime and white cement, so call them in advance. The amounts needed will vary according to the size of the house, the depth of the cut-back joint and your abilities as a pointer. The rubblestone example would require about two single-axle truckloads of sand, twenty-five bags of hydrated lime and twelve bags of white Portland cement. Consult Chapter 9 for estimating your needs.

Before erecting your rented or borrowed scaffolding, determine which plants will be affected by the scaffolding itself, possible cleaning procedures, chipping of old mortar and excess new mortar. Cleaning (see page 169) may mean that plants should be removed to protect them from chemicals. Even simple chipping and repointing of a wall thirty-two feet long by fifteen feet high will leave approximately a foot of debris at the base.

Put the scaffolding up along one entire wall. Start with the rear wall where initial mistakes won't be so obvious. On our thirty-two foot wall we will need four sections in length and two sections in height. Make sure you have adequate planking and safety rails. Position scaffold ladders at convenient points.

Garbed in hard hat, goggles (eye injuries are the most common in this kind of job), work boots and gloves start chipping the joints out beginning at the top of the wall. The center of the joint should be struck, allowing the mortar to fall away from the edges of the more solid material. If an incorrect, harder mortar has been used in the past, this striking method will not damage the edges. The joints should be cut back at least half an inch to accept the new material. Remember you are removing the mortar, not chipping the material of the wall.

After about fifteen minutes your hands will be ready to fall off. Surely there must be an easier way to do this. There is. A chipping tool called a scaler (see Chapter 9) is an air-powered device used for this purpose. It may be rented along with the necessary compressor. The scaler is fast and effective, but particular care must be taken not to allow the tool to run away from you and damage the edges. When working on brick, this device should only be used

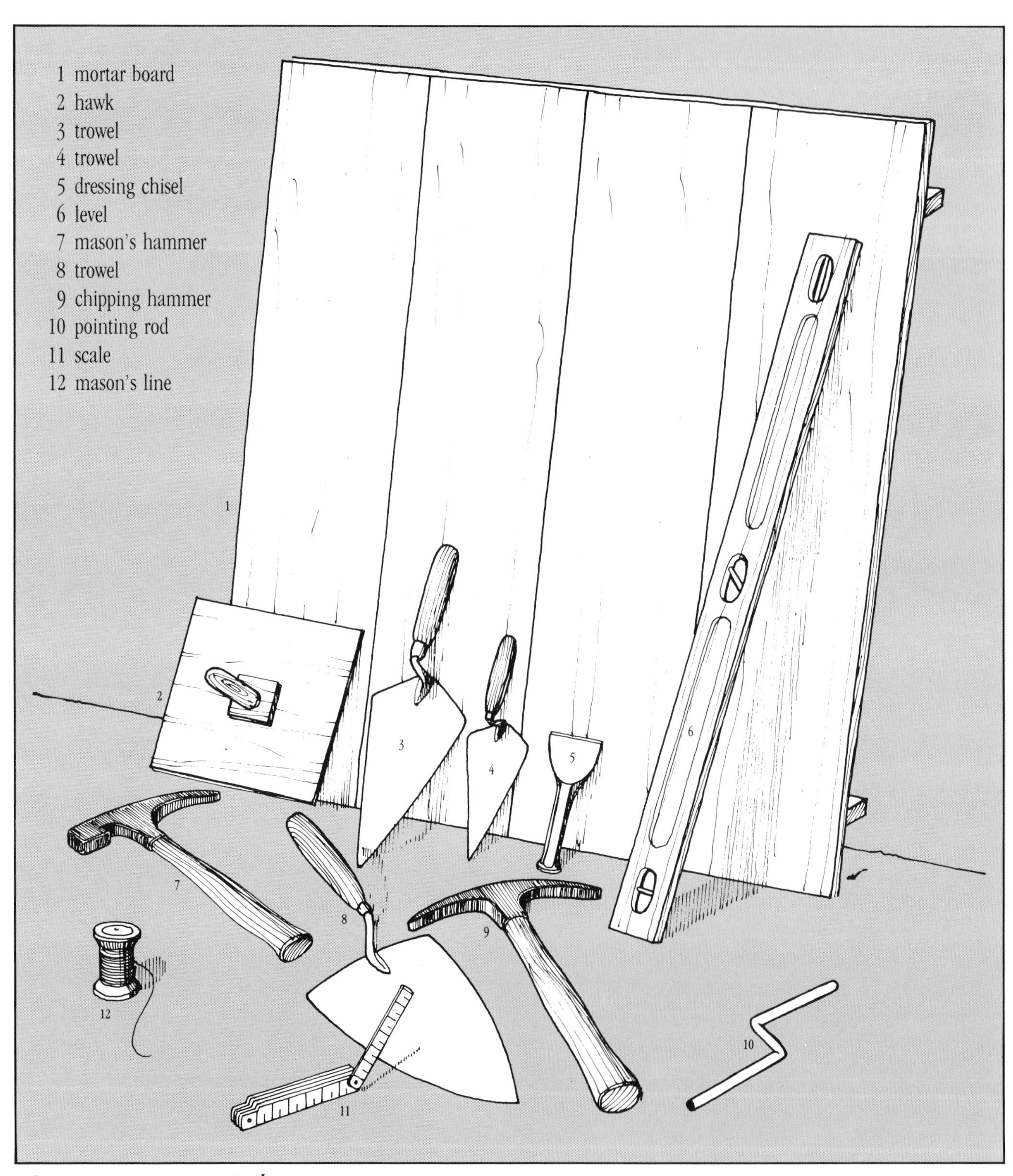

Contemporary masonry tools

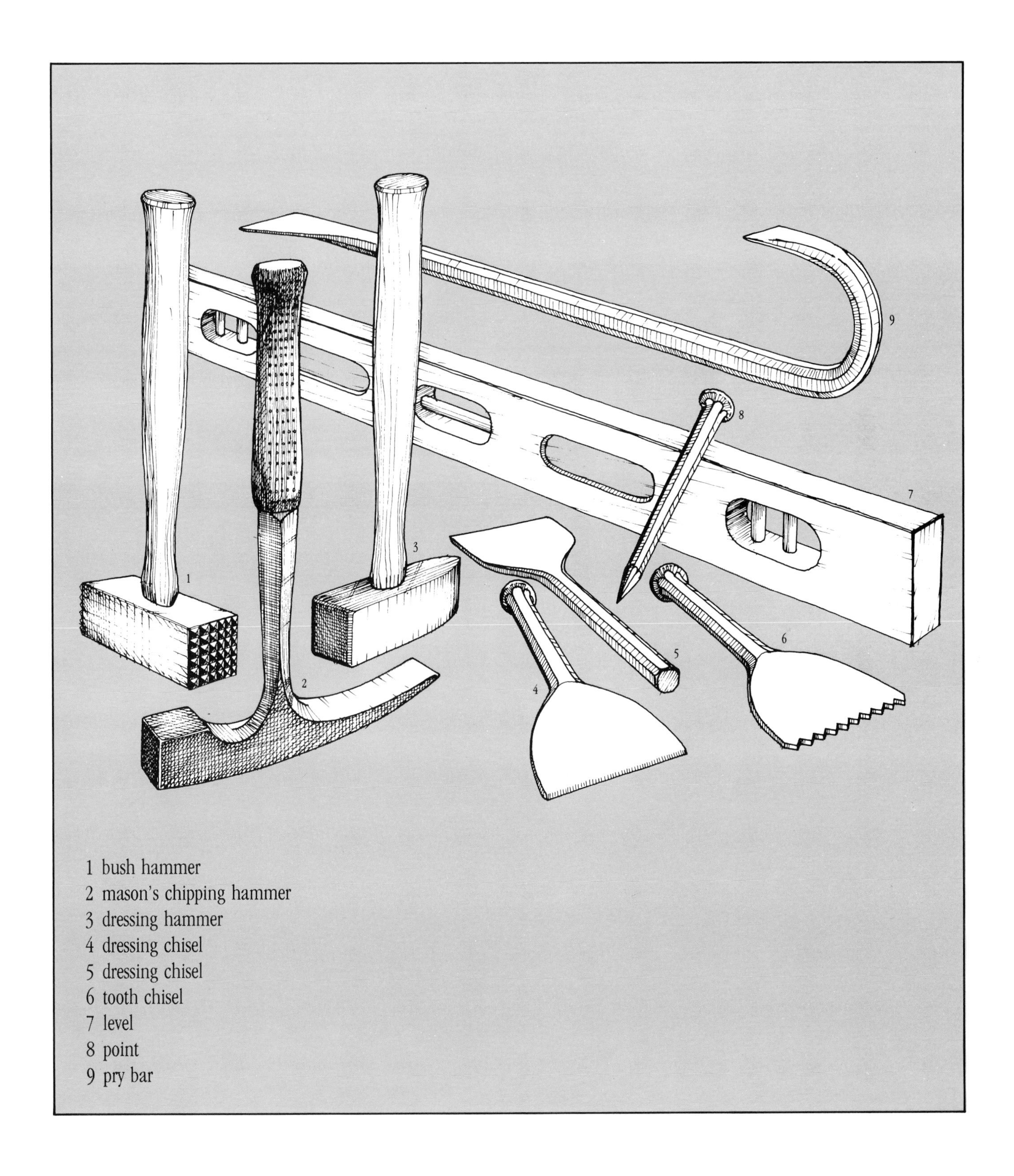
1
2
3
4
5
6
7
8
9
1 bush hammer
2 mason's chipping hammer
3 dressing hammer
4 dressing chisel
5 dressing chisel
6 tooth chisel
7 level
8 point
9 pry bar

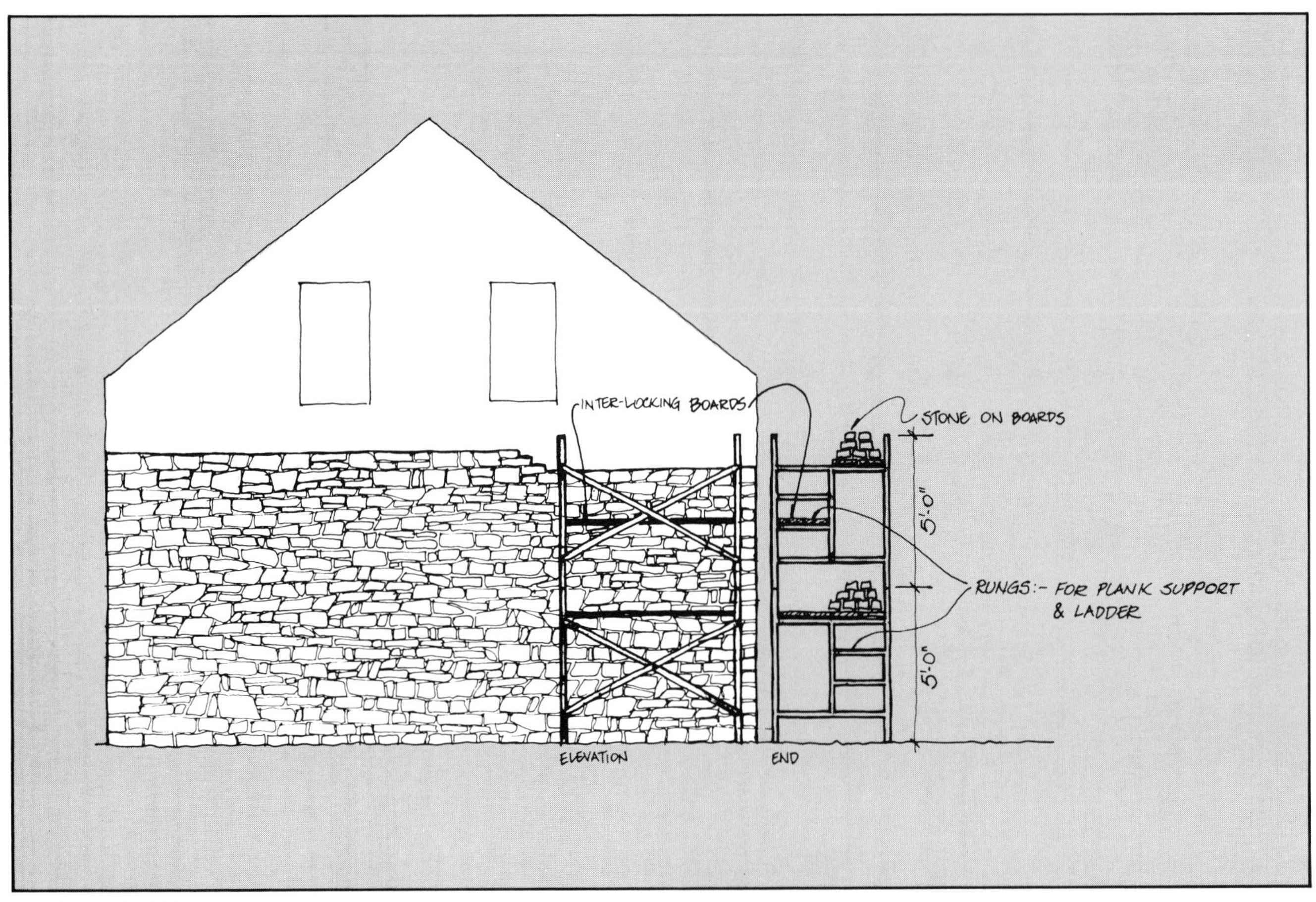

Safe scaffolding

OPPOSITE, ABOVE

When repointing a brick wall, techniques and materials that are compatible with traditional methods must be used. The joints here have been cut back by hand. The finished pointing should resemble the left-hand portion of the wall, where mortar has been applied to the joint and a white lead highlight added to it. This is called tuckpointing.

OPPOSITE, BELOW LEFT

This uncoursed rubblestone house has had the joints cut back and lightly sandblasted. The blasting was done to remove Portland cement staining from a previous, incorrect pointing job.

OPPOSITE, BELOW RIGHT

Here the mason is rubbing down the freshly pointed joints in this rubblestone wall to give a slightly indented profile.

when entire units are being replaced, otherwise it will tend to shatter the edges around the much narrower joints. On the other hand, a scaler is most useful on rubblestone that has been incorrectly pointed using straight Portland cement. Rental costs demand that the job be done with some haste or the economic advantage of doing it yourself will be nullified.

If you are repointing brick, you are not necessarily confined to handwork. It is possible to rent a rotary power saw that will actually cut the joint. The blade should be slightly thinner than the joint. Extreme care should be exercised, because the tool can be difficult to control, and overruns will score the brick.

No matter what chipping method is used, work across the wall and down. Once the entire wall is chipped, clean away debris from the scaffold and the base of the wall. It is easier to do now than when fresh mortar has been dropped on it. Rinse off the wall using an ordinary garden hose with a nozzle. Make sure the joints are free from debris.

To prepare for the actual repointing process, make sure the materials are stacked in convenient areas, ensure an adequate supply of potable water, and clean out your trough or mortar mixer. For our model building, the mortar should be mixed according to the following recipe:

12 parts mortar sand
3 parts hydrated lime
1 part white Portland cement

Mix only as much as you can easily use in fifteen to thirty minutes. Don't throw the dry materials together with a shovel, measure them out in the rubber pails. Blend them thoroughly and then add water, a bit at a time. The finished consistency should be moist but stiff — similar to soft brown sugar. Re-wetting dried mortar is not effective.

If the joints to be filled are larger than an inch and a half, you may need to do the job in two coats. The first coat can be a bit wetter than the second.

Again, start at the top of the wall and work your way across and down. Buckets of mortar can be hoisted to the top of the scaffold with a simple rope and tackle. However, if this is a one or two person job, it is probably more effective to bring all the mortar up at the same time to avoid wasting time.

The tools needed at this point are a pointing trowel of the appropriate width and a mortarboard. The mortarboard may be as fancy as an aluminum plasterer's "hawk" or as simple as a piece of scrap board with a handle attached. Place some mortar on the board and hold it parallel to the lower edge of the joint. Force the mortar into the joint with the trowel. The mortar should be flush with the face of the stone. After pointing a small section, step back and view your work. It will probably look a mess. Follow the same procedure throughout the wall.

The second stage of pointing will improve matters; this is called tooling or joint finishing. After the mortar has set up and dried (about two hours depending on the weather), take an unprinted burlap sack and rub the joint down until a slightly concave profile is achieved. (The ink in printed sacking runs when wet and will stain.) During the rubbing process excess mortar will fall away, leaving clean, slightly indented joints with unblemished faces. This concave style is the best joint for most masonry structures; however, local vernacular tooling may vary.

When pointing a large wall that may take more than one day to complete, leave off in an irregular fashion. When you return next day, the slight variations in color and texture of the mortar will not be so visible if the completed joints have been staggered.

Consult your local weather office as pointing progresses. A sudden downpour can pit the joints and make them less than useful, but a small sprinkle probably won't do any harm. In extremely hot, fair weather, let the mortar dry rather slowly by hanging wet sacking over but not touching the completed wall. The final step is the least pleasant. Clean up the

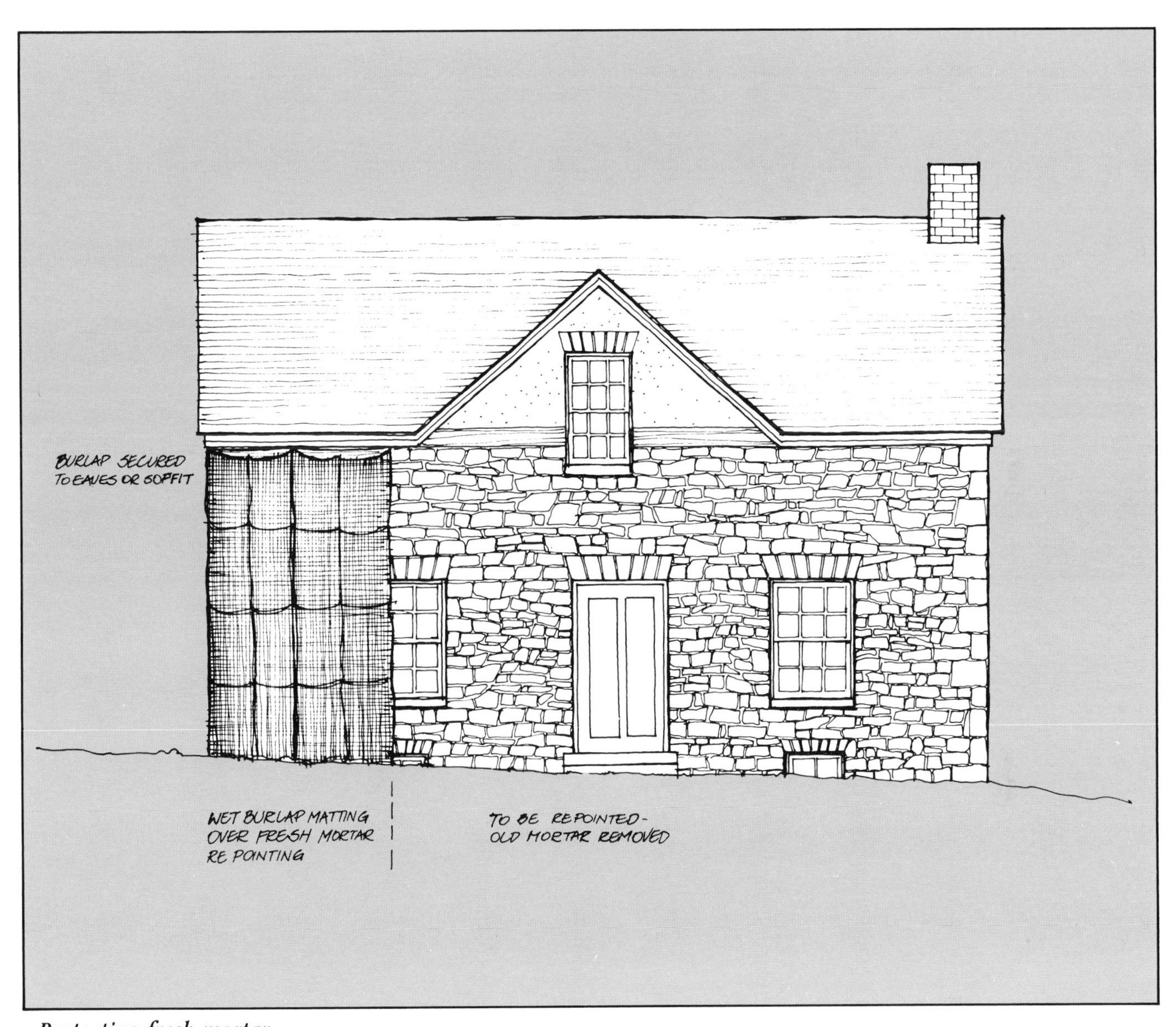

Protecting fresh mortar

Fine joints in ashlar should be repointed with great care so as not to mar the joint edges.

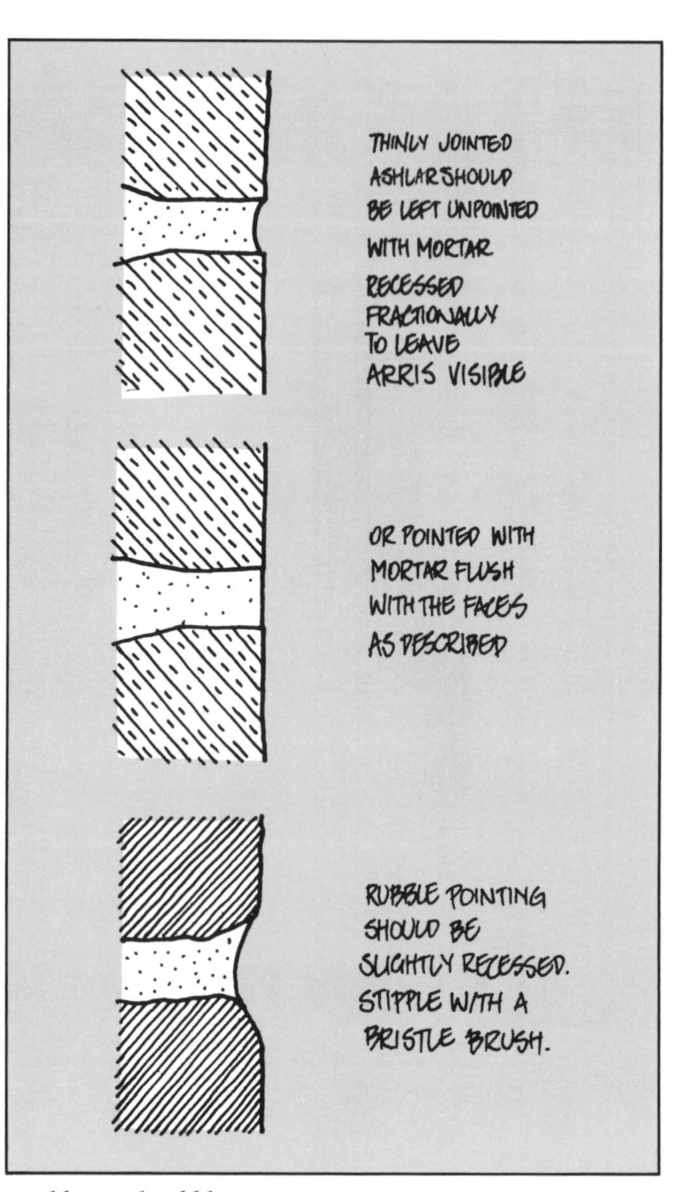

Ashlar and rubblestone joints

rather large quantity of mortar that has been rubbed from the joints and is now deposited at the base of the wall.

Most stone and brickwork can be pointed in this fashion. The one exception is ashlar or cut stone walls with very narrow joints. Repointing in this case is often unnecessary and may cause more harm than good. Never widen an ashlar joint to accept repointing. If it is absolutely necessary, cut the joints back less than an eighth of an inch, use the same type of mortar described earlier and finish the joints flush with the stone. Tape either side of the joint to ensure a clean, straight line of mortar.

Although mortar mixes and joint tooling may vary according to regional or vernacular styles, the basic instructions given here will make it possible to repoint any wall. A good repointing job will last fifty to one hundred years. Many historical ones have lasted much longer. Remember, incorrect materials or inadequate craftsmanship will cost more in the long run.

Coatings

Modern masonry coatings, both waterproof and water-repellent, are frequently recommended for their preservative abilities. Though their benefits sound advantageous, they should be approached with extreme caution or ignored completely. Traditionally, coatings were applied for either visual or maintenance reasons. They usually repelled water to some extent, but they also breathed, allowing interior moisture to escape. One of these, stucco, is described here, and other recipes are found in Chapter 9.

It is often desirable to repaint a masonry building for aesthetic or maintenance reasons. To guarantee good bonding, the surface should be free from dirt and alkalis. Latex paints are recommended.

Actual sealers that prohibit the intrusion of moisture have been known since early times. A combination of liquid hot wax and paraffin was one of these. Such coatings work only as long as all water is repelled. If moisture enters the wall, it will take the path of least resistance, escaping inwardly to damage interior decor and furnishings. The use of coatings in any form must be examined with economics and general wearability in mind. It may be relatively cheap to paint the exterior of the house, and for five years have it look quite chic, but repainting will be necessary within five to eight years. Cleaning may be more expensive initially, but if done carefully, it may provide desirable visual charm as well as maintenance-free service for your lifetime.

Stucco, pargeting or parging was one type of traditional coating that was widely used. The process involved covering exterior walls with a thin coat of plaster-like material. (Chapter 9 lists various recipes.) Stucco was primarily a statement of fashion, and it was made up of a mortar mix similar to that described under repointing but with the addition of animal hair for increased binding strength. The mixing of materials can be done by hand, but unlike repointing the amount of materials used in half an hour will be fairly substantial. For this reason, it is advisable to rent a mortar mixer, not a cement mixer which is larger and has a different shape.

Most coatings of this type were, and still are, made up of three coats: scratch, brown and finish, although occasionally only two coats were used; one thick base and one thinner finish coat. The materials, techniques and finishes are quite similar to interior plastering, which is covered in Chapter 6. If you decide to recreate a traditional stucco finish, use the same recipe given for pointing mortar but buy finer sand. Animal hair can still be obtained from stockyards or tanning factories.

It is imperative that no lumps remain in the mix, because they will mar and score the surface. You will need many of the same tools listed under repointing, with the addition of a rectangular "laying-on" trowel. It is possible to economize in many areas, but this is not one of them. Trowels vary widely in price, but in my experience the cheap variety is flim-

ABOVE

Roughcast or harling provides a sound, historically correct coating for this mid-nineteenth-century rubblestone house.

LEFT

Harling applied to riven or handsplit lath as an exterior coating on an early nineteenth-century house.

sy, difficult to work with and lacks the necessary spring in the blade. Invest in the best trowel you can find, one made by a reputable company like Marshalltown, for example. It will have many uses, not the least of which is interior plastering.

The first stucco coat is basically slapped on, so a few lumps won't be too noticeable, but the third coat must be lump-free. The mixture should be firm enough to stick to the surface but slide off the trowel. A good wrist action is important here; a flick of the wrist should free the mortar from the trowel and adhere it to the surface where it can be spread evenly.

Reparging is most often a repair operation, rather than the recoating of an entire wall or walls. Make sure the area is clean and free of debris. Gypsum, as found in contemporary plaster mixes, causes an adverse chemical reaction with stucco, so their use together should be avoided.

Test the porosity of the wall by spraying it with a hose. If the water runs off, it will not accept stucco. The wall should be roughened by chipping it lightly with a chipping hammer and then cleaned further. If the water is sucked up very quickly, make sure the wall is lightly wetted before the first coat is applied. Cracks and poor adhesion will be the result if the moisture in the stucco is drawn into the wall surface.

This coat is called the scratch coat, because the surface should be scored after it has set but not dried. It should be about one-quarter of an inch thick. Scratch the surface randomly in long lines, but don't scrape it off the wall. The idea is to provide "keys" for the second coat, not deep gouges. The brown coat is applied relatively soon (four to eight hours) after the first. The first coat should be firm but not completely dry. Moisten the scratch coat with a fine mist from a garden hose if it seems too dry.

The second coat should be one-quarter to three-eighths of an inch thick and applied with your laying-on trowel. This application should fill any voids or identations, and although the surface should be on an even plane, it should also be irregular enough to accept a third coat. The brown coat should not dry too quickly. The burlap mat described under repointing should be employed if the weather is hot or you are working on a south-facing wall. Cover the wall for two or three days to allow the material to cure slowly. If it rains after the first few hours, don't worry. Rain will not damage the surface appreciably.

Nine to ten days curing time should elapse between coats two and three. The finish coat should measure one-eighth to one-quarter of an inch thick. Trowel in the same manner as the second coat, but try for an even plane and smooth surface.

Some stucco finishes on rough rubblestone buildings were tooled to resemble ashlar or cut stone. If you are repairing a wall finished in this manner, or one employing English roughcast or harling, allow the third coat to dry to a uniform firmness before beginning. When tooling an entire wall, it is advisable to work the pattern out beforehand on paper, including sill and lintel details. An early description of the process will give the best explanation:

> The object of covering the outside of the walls of cottages with cement is generally to imitate stone. In this imitation, care must be taken that the lines drawn do not represent stones of too large a size; that the shapes of the stones at the corners, and for the lintels and sills of doors and windows, be suited to their situations and uses; and that, in the regular courses, the joints alternate and show bond properly, as in regularly built stone buildings.[1]

English roughcast stucco, on the other hand, employed a much different type of finish; it was done by:

> dashing the surface of the plaster, after being newly laid on, with clean gravel, pebbles, broken stones of any kind, broken earthenware, scoria, spars, burnt clay, or other materials of the like description, sifted or screened, so as to be of a uniform size. The effect of surfaces of this kind is good, and the process admits of producing very great variety in the external appearance of cottages. By

being forcibly thrown against the moist plaster, they penetrate into it, and render it very firm and durable. Sometimes, instead of the stones or other matters, being broken to a small and uniform sized gravel, they are pounded into a coarse sand, and this is dashed against the moist mortar. The effect is pleasing, but the strength and durability are not so great as in the other mode. In using small stones or gravel, it is desirable, for the sake of effect, previously to render the moist plaster as nearly as possible of the same colour as that of the materials to be thrown against it. It is also desirable that all corners, sills, lintels, and, in short, all vertical and horizontal bond, should be tinted of the same colour as the roughcasting.[2]

Another type of roughcasting, or harling as it was called in Scotland, was employed throughout North America. The final effect was somewhat like English roughcast, but the process was markedly different:

Plaster the wall over with lime and hair mortar; when this is dry, add another coat of the same material, laid on as smoothly and evenly as possible. As far as this coat is finished, a second workman follows the other, with a pail of roughcast, which he throws on the new plastering. The materials for roughcasting are composed of fine gravel, reduced to a uniform size by sifting or screening, and with the earth washed cleanly out of it; this gravel is then mixed with pure newly slaked lime and water, till the whole is of the consistency of a semi-fluid; it is then forcibly thrown, or rather splashed, upon the wall with a large trowel, which the plasterer holds in his right hand, while in his left he has a common white-wash brush. With the former he dashes on the roughcast, and with the latter, which he dips into the roughcast, he brushes and colours the mortar and roughcast that he has laid on, so as to make them, when finished and dry, appear of the same colour throughout.[3]

No matter what type of finish is applied to the stucco, the wall should be kept moist and burlapped for two or three days before exposing it to the sun. As in pointing, a most horrendous mess will surround the area, and care should be exercised to remove plants that will be adversely affected.

Although the thicknesses of the various stucco coats have been specified, when repairing a wall it is always best to try to imitate the original method. By chipping a small section away from an edge, a profile will reveal thin, slight color variations in the three coats. Wherever possible exact historic measurements should be followed, using compatible materials and techniques.

Masonry Cleaning

Although the cleaning of masonry is usually left to a professional, the owner of a brick or stone home should be familiar with the various methods, their benefits and pitfalls. It is a labor-intensive, dirty and sometimes hazardous operation, but it is possible to rent the equipment and do your own work.

Before any cleaning is done, ask yourself why it is necessary. Perhaps it is cosmetic, and a wash-up will make the building look sparkling new again. It may be a public statement of one's intention to restore or upgrade the property. But is it historically correct? Will the building be damaged?

The history of masonry construction in North America, primarily of brick buildings, gives ample evidence that washes or coatings were applied for aesthetic and protective reasons. Porous, handmade brick is very susceptible to deterioration when exposed to the harsh climates and wide temperature swings of the northern United States and Canada. Protective coatings were regularly applied to slow the disintegration process.

"Dirt may be defined as finely divided solids held together by small amounts of organic material."[4] When making a decision about cleaning, a good rule of thumb is that unless the dirt is causing some form of damage or appreciably detracting from the historical impact of the building, it is best left alone. A thin coating of time-accumulated particles does act as a protective barrier on brick and stone. Removal may do more harm than good, and it may form part of the visual charm of an old house.

However, a heavy buildup of dirt may hold excessive moisture, which, in turn, causes deterioration.

In heavy industrial areas, acid rain, as well as other airborne pollutants, will adhere to this surface and lead to major physical change in the masonry structure. Other common menaces to the masonry surface may occur in the form of algaes, lichens, ivy or creepers. The branches of mature trees may lash and stain the walls. The evil pigeon and its feathered friends are renowned for their good taste in domiciles; their aim is unerring, always the elegant facade of your house. Droppings, unsightly as they are in themselves, cause a chemical reaction that will pit brick or stone. Excessive salt staining will lead to material breakdown over time. Occasionally, metal bonding or tie rods will rust due to natural weathering, leaving long rust stains. Construction negligence may have deposited tar grease or cement stains that detract from the overall appearance of the house. Depending upon the degree of and the potential for actual damage, these may be sound reasons for cleaning.

The mid-twentieth-century fixation on things natural started with the rampant sanding of pre-1900 floors. Now the strippers are turning their talents to exterior masonry walls, where their efforts may do even more damage. Brick, in particular, was probably painted for a reason; it is difficult to know just how long masonry buildings will last that have been wantonly stripped of their protective coatings.

The three main forms of cleaning are water, chemical and mechanical. Before choosing one of these, ask yourself the following questions. What will I damage? Particularly in chemical cleaning, this may apply to the building itself, plant life and animals. In the case of paint, why was it applied in the first place — for historic or maintenance reasons? Will airborne chemicals or abrasives damage nearby properties? The windshield of our truck had to be replaced due to the etching caused by chemical overspray from an adjacent site. Will woodwork, glass or general paintwork be adversely affected? What about personal safety? Abrasives cause excessive dust, often containing silica. Chemicals, both liquid and solid, may pose health hazards not only for the operator but also for the entire neighborhood. Negligence leads to accidents. Burns from steam, disfigurement from sandblasting — these are serious matters and should be carefully weighed.

Once you are convinced that cleaning is necessary and the risks minimal, the first step in the actual process is to prepare a number of test patches, preferably in non-conspicuous areas. It is not unusual to combine different types of cleaning on one job, because various grades of brick or stone may respond better or worse to a specific method. The test patches should be made on each wall or area to be cleaned. Ideally, the tests should be left for one year so that the effects of weathering can be examined. If no apparent problems occur, it is safe to proceed.

If the job is to be sub-contracted, it is imperative that the company chosen has a high standard of workmanship. Recent samples of the company's work should be inspected, or a professional preservationist should be consulted about its abilities. The very real difference between cleaning a modern and a historic building is probably best illustrated by the following example. High-pressure water cleaning may be considered as anything over 5,000 psi (pounds per square inch). In the case of a modern industrial building, the pressure might be as high as 15,000 psi. By contrast, a pressure of 10,000 psi used on an 1830 brick house would, no doubt, explode the surface. At a distance of one foot from a historic surface, a pressure of between 20 and 100 psi should be employed. This may vary because of the original materials used, but the point is to start weak and build up.

After determining the best system or systems to use on your house, and after making very sure that the contractor understands how to clean historic buildings, be certain that the contractor has adequate permits and liability insurance. It is a matter of courtesy to inform neighbors what will be taking place.

If you plan to do the job yourself, scaffolding must be rented along with the appropriate equipment. Protective equipment is essential, including hard hat, gloves, goggles, rubber slicker and pants, and in some cases an appropriate mask. Estimating, equipment types and problem stains are discussed in Chapter 9.

Moisture in Basements

Some early North American masonry buildings literally grew out of the ground. The first habitations in New England, for example, were sometimes no more than holes in the ground or dugouts in the sides of hills:

> Those . . . who have no means to build farmhouses . . . dig a square pit in the ground, cellar fashion, 6 or 7 feet deep, as long and as broad as they think proper, case the earth inside with wood all around the wall, and line the wood with bark of trees or something else to prevent the caving in of the earth, floor this cellar with plank and winscott it overhead for a ceiling, raise a roof of spars clear up and cover the spars with the bark or green sods, so that they can live dry and warm in these homes with their entire families, two, three or four years.[5]

Often, such crude homes became the bases for new houses constructed above grade, and they sometimes continued to function as kitchen areas. The present-day owner of a period home may be fortunate enough to find such a kitchen, complete with large cooking fireplace and built-in cupboards. But it is much more likely that all you will find in the basement is a host of problems.

Far from "dry and warm," the average period basement is so damp that the space is unusable. To correct the problem, the source must be found. Moisture may be due to poor runoff, usually because of an inadequate drainage system. Over the years, both surface and subterranean drainage have deteriorated, become overgrown by vegetation or been redirected because of ground level buildup. A classic example is when early eavestroughing has become clogged or has rotted through. The water overflows, lands on a driveway that has been paved but badly joined to the foundation, pours between the driveway and foundation, seeps through the mortar, and ends up in the basement.

Soil conditions themselves often lead to moisture problems. Clay near a foundation will result in extremely slow runoff. The clay actually acts as a barrier, which holds water and leads to the next major problem, condensation. For example, if a basement is seventy-five degrees Fahrenheit (24°C) in the summer and the water-logged clay remains at forty degrees Fahrenheit (4°C), the result is sweating, dampness, condensation.

Moisture control should start on the roof. When rain or snow hit the roof, where does the water go? Wherever possible guttering or eavestroughing should be installed so that the down-pipes direct water downgrade from the foundation. Clean gutters annually; clogged they're useless. Once the heavy debris has been removed, flush them out with a garden hose. While up there, make sure that all fastenings are secure.

An even more serious problem occurs in early houses with eaves or dormers that have been insulated. The areas where eaves and wall join, or dormer and wall meet, cannot be adequately insulated; warm air leaks out and ice dams form in winter. About the only solution is to install a heater cable in the culprit area. The cable plugs into an external outlet. When the temperature dips below freezing, a thermostat trips on the heat and the ice melts.

Changes in landscape are probably the single major cause of basement water penetration. Ideally, all foundations should have the grade sloping gently away from them. Major plantings may make this impossible, but if there is a severe penetration problem the only real solution is to excavate the entire foundation and put in weeping tiles. This is a large undertaking, but it is simple. With judicious care it can be accomplished by the owner-preservationist.

The whole earth philosophy in the nineteenth century: an uncoursed rubblestone house built into the side of a mound.

First of all, any vegetation of value should be removed. If this is to be a long-term project, it is often better to replant in another location for the season. Bear in mind that in a trench four feet wide and approximately eight feet deep there is a lot of earth. Make sure you have allowed enough space for the dirt and for the backhoe to maneuver.

When the backhoe is hired, instruct the operator to excavate approximately six inches to one foot away from the foundation wall. It is possible that the wall will not be sound. The earth left by the backhoe will usually fall away from the foundation by itself, but if it doesn't, you can usually shovel it out as you go.

Footings were usually no more than rubblestone pads extending about one foot on each side of the foundation wall. Excavate down to the footings around the perimeter of the house, hose down the walls and make sure they are free of earth and debris. As in repointing (see pages 64 to 73), joints should be cut back until the mortar is sound. Because of constant water leaching, it may be necessary to dismantle and rebuild a section of wall (see Chapter 4). All joints should be repointed using traditional materials.

Personally, I enjoy this kind of excavation, because one can never tell what will be unearthed: old bottles, hardware, coins, pottery shards and so on. One old house owner with keen eyes was able to salvage a circa 1830 fireplace trivet before it was smashed by the backhoe. Once my archaeological impulses have been satisfied, I tend to overcompensate for potential future problems while the foundation is exposed. The usual practice is to brush heavy tar over the walls after repointing. To be even more safe, however, I ordinarily point the foundation and then cover the entire wall with a parging of Sealbond. The parg-

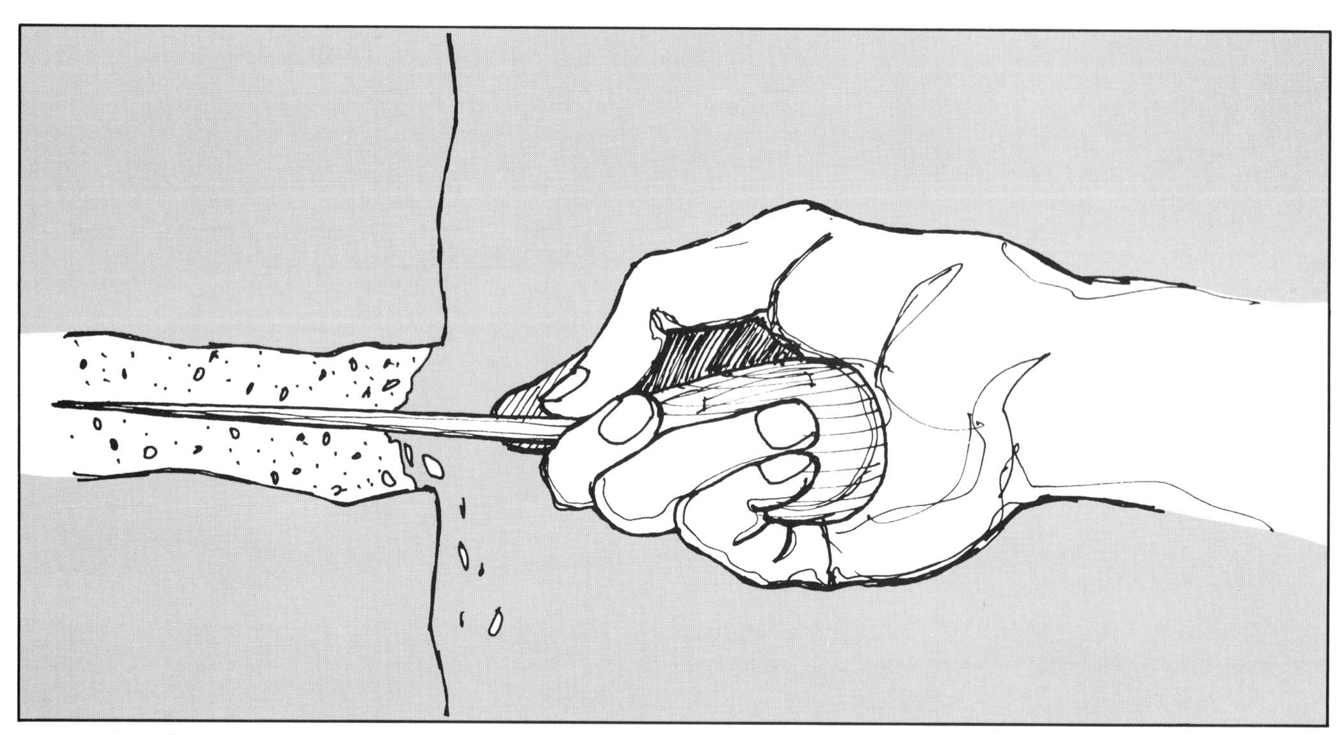

Ice pick probe

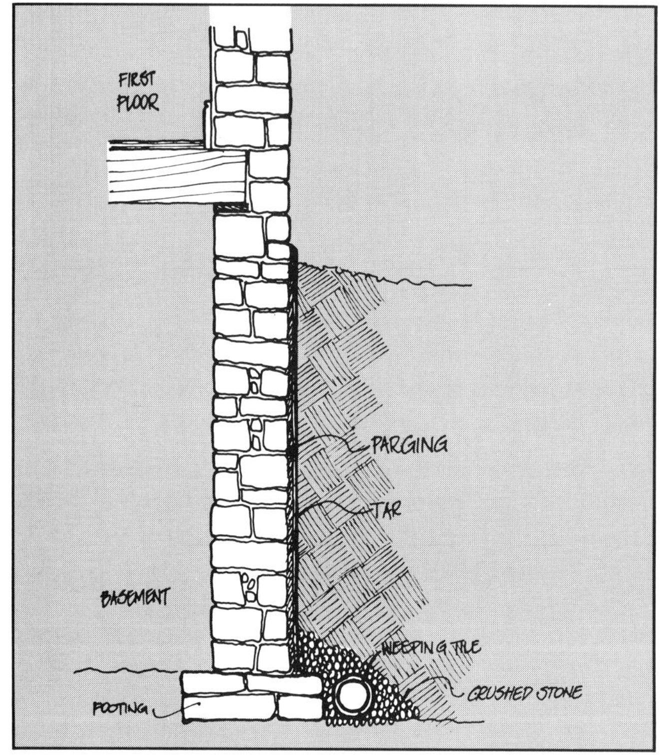

Well-drained basement wall

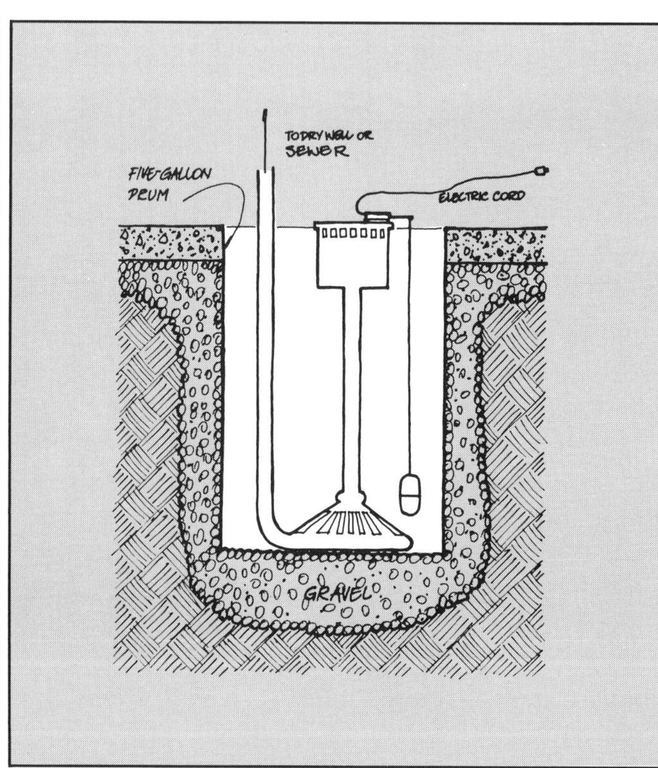

Typical sump pump installation

ing should be flat, smooth and finished off with a roller-applied coat of warm tar. While the tar is still relatively tacky, I apply two-inch polystyrene insulation sheets to the wall. The tar acts as an adhesive. The sheets should extend from the footing to slightly above ground level. At the end of this part of the process, the foundation is sound, waterproofed and insulated.

You may have noticed that the basement has been relatively dry since excavation. This is because the moisture-holding materials have been removed, and the entire foundation system has had a chance to dry out. But we still have to deal with the problem of excess moisture. Lay a deep bed of crushed stone or aggregate to slightly above the footings, place a four-inch drainage tile on it, and cover with more stone. In some cases, such as in areas of high clay content, the stone should be continued up to grade level. Fill in with crushed stone approximately one foot out from the foundation walls, and then backfill with the earth from the original trench. In urban areas the tiles should be directed into a storm sewer; in the country use a dry well or lower ground.

To give you some idea of the advantages of installing weeping tiles, the following permeability table compares crushed stone to other materials.

Material	*Permeability in feet per day*	*Flow in gallons per minute*
Crushed stone	30,000	45
Clean pea gravel	1,000	10
Fine sand	10	1/10
Silt clay	1/1,000	1/100,000

There are a few things to bear in mind when using a backhoe. It is usually rented on an hourly basis, with a fixed float charge for transportation. Therefore, it is more economical to do all the excavation at one time and all the backfilling at another, so long as the foundation is reasonably sound. Have all the sand and materials for repointing and parging delivered on the same day as excavation. The hoe can easily distribute the load to three or four mixing locations. When the machine returns to backfill, coordinate the delivery of the crushed stone. Let the backhoe move the stone, while you place the tiles. Anyone who has wheeled 200 loads of mortar around a foundation or shoveled tons of stone will immediately see the brilliance of such a scheme. What would take three experienced people two weeks will require two good workers, one backhoe and one Saturday.

It may seem like overkill, but once the exterior of the foundation is repaired, the interior should be inspected as well. A simple method to test interior joints is by probing them randomly with an ice pick. If there is resistance after about a quarter of an inch, the mortar is acceptable. If the pick sinks to the handle, start repointing.

Even in a basement that is relatively sound and dry but has some moisture problems — and nearly all rural basements do — a simple sump pump may be the answer. Choose a site for the sump pit at the lowest level of the basement, break a hole in the existing floor and dig a hole ten inches in diameter and about two feet deep. Cut the bottom out of a five-gallon can, set the can in the hole, mortar around the top, put some crushed stone in the bottom of the hole and set the pump in place. The sump line should be routed through the foundation wall or out a basement window and directed away from the house so the water will not run back in again.

Many early houses were not constructed with anything approaching a full basement. Today's lifestyle often requires such a space for mechanical systems, workshop and so on. There is no easy way to enlarge the basement under an existing house, but several methods are explored in Chapter 9.

CHAPTER FOUR

Building with Stone and Brick

The interior side of a new, uncoursed rubblestone wall built in a traditional manner. Above the wooden lintel is a dook or furring strip for nailing on framing.

A COMMON PROBLEM for the contemporary old house owner is the enlargement of traditional space to accommodate modern lifestyles and mechanical systems. Offices, laundry areas, bathrooms, heating units — all require larger amounts of space than one ordinarily finds in the historic home. The immediate solution is to build an addition to the original structure. In considering an addition, compatibility must be the chief concern. It is possible, of course, to build a wooden addition copying clapboard or board and batten methods, but it is much more preferable to use the original material.

This chapter is devoted to construction in stone and brick in the traditional manner but utilizing twentieth-century building alternatives. The instructions given here will allow you to build both an addition or an entire house. Although you may not be interested in constructing either one, understanding fundamental building techniques is a useful tool for anyone engaged in the preservation of period houses.

No matter what the material chosen for an addition, step number one is to devise a building plan. Attention must be paid to the proportions of the original structure, so that the addition becomes neither unwieldly nor obtrusive. Ethnic or vernacular details of trim, bonding, pointing, window treatments and so on should all be taken into account. Step number two is to acquire the necessary materials. Quarried rubblestone can still be bought and delivered to your site, but the subject here is preservation. It will probably be very possible to acquire collapsed stone houses or derelict barn foundations in your area for a much more reasonable fee. The advantage is that the original mason has already dressed the stone for you. Brick, on the other hand, can also be bought new, but the texture and absolute uniformity of modern units will not blend with the main structure. Wrecking companies are an easily accessible source for old brick; however, there is a difference between old brick and the antique variety. The best thing to do is to take an original brick from your property and try to match it in color and texture.

Although reclaimed stone and brick are usually cheaper than new materials, trucking costs and in some cases the use of a front-end loader at the demolition or building site should be considered. Most stone buildings were constructed with soft lime mortar, and they will come apart relatively easily. The mortar will fall away from individual stones making reuse fairly simple. Recycling your own brick may be possible, but pre-1900 units are quite soft, the breakage rate is enormous, and your time would probably be better spent finding a dealer who specializes in period materials.

The Stone Addition

It is important to remember when building with stone that the early footing was little more than a pad extending eight to twelve inches on either side of the actual foundation wall. The same general principle can be followed for a new footing. If a two-foot-thick wall is to be erected, bear in mind that a contemporary concrete or block basement wall is only nine or twelve inches thick. Therefore, unless you

OPPOSITE, ABOVE LEFT
New construction often employs a combination of traditional and contemporary methods. Here, a concrete foundation has been poured for the addition and garage.

OPPOSITE, ABOVE RIGHT
Two-by-four framing and friction-fit insulation batts have been installed on the interior walls. The void will accept a new fireplace.

OPPOSITE, BELOW
The finished restoration with its harmonious addition.

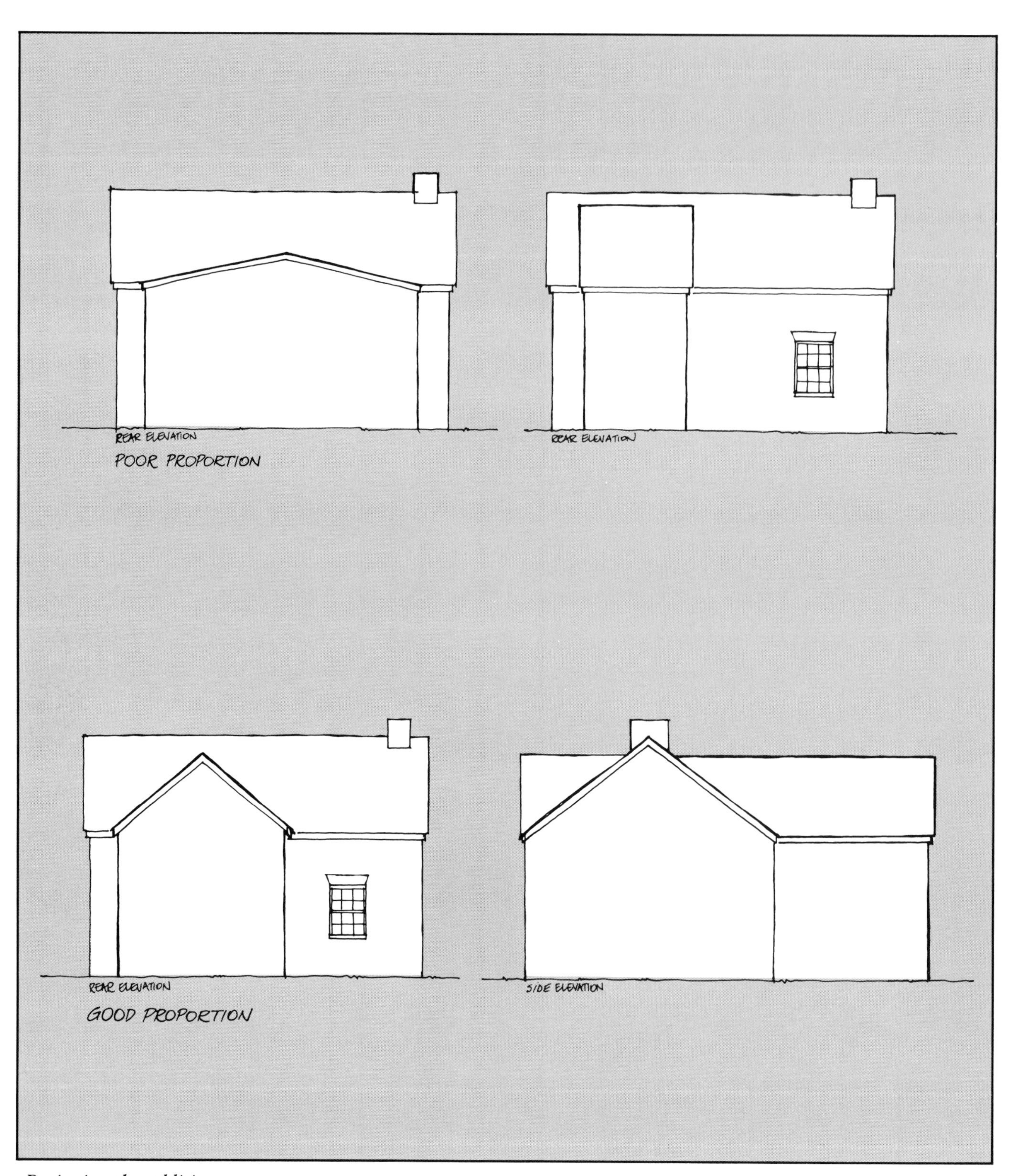

Designing the addition

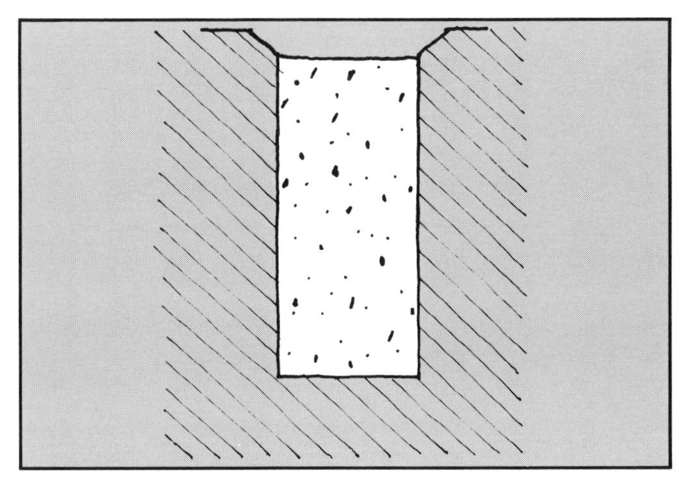

A slipform foundation

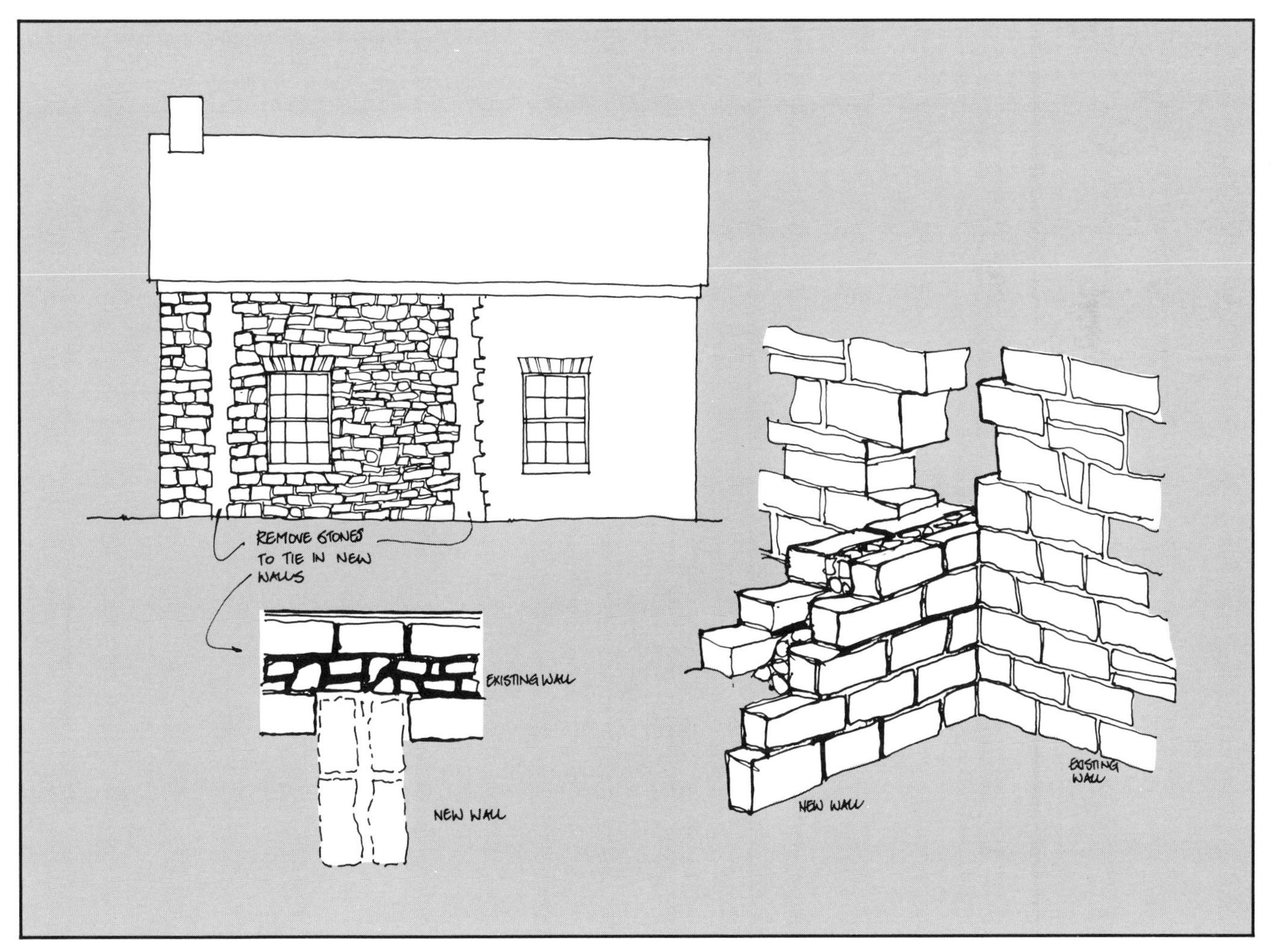

Tying in the stone addition

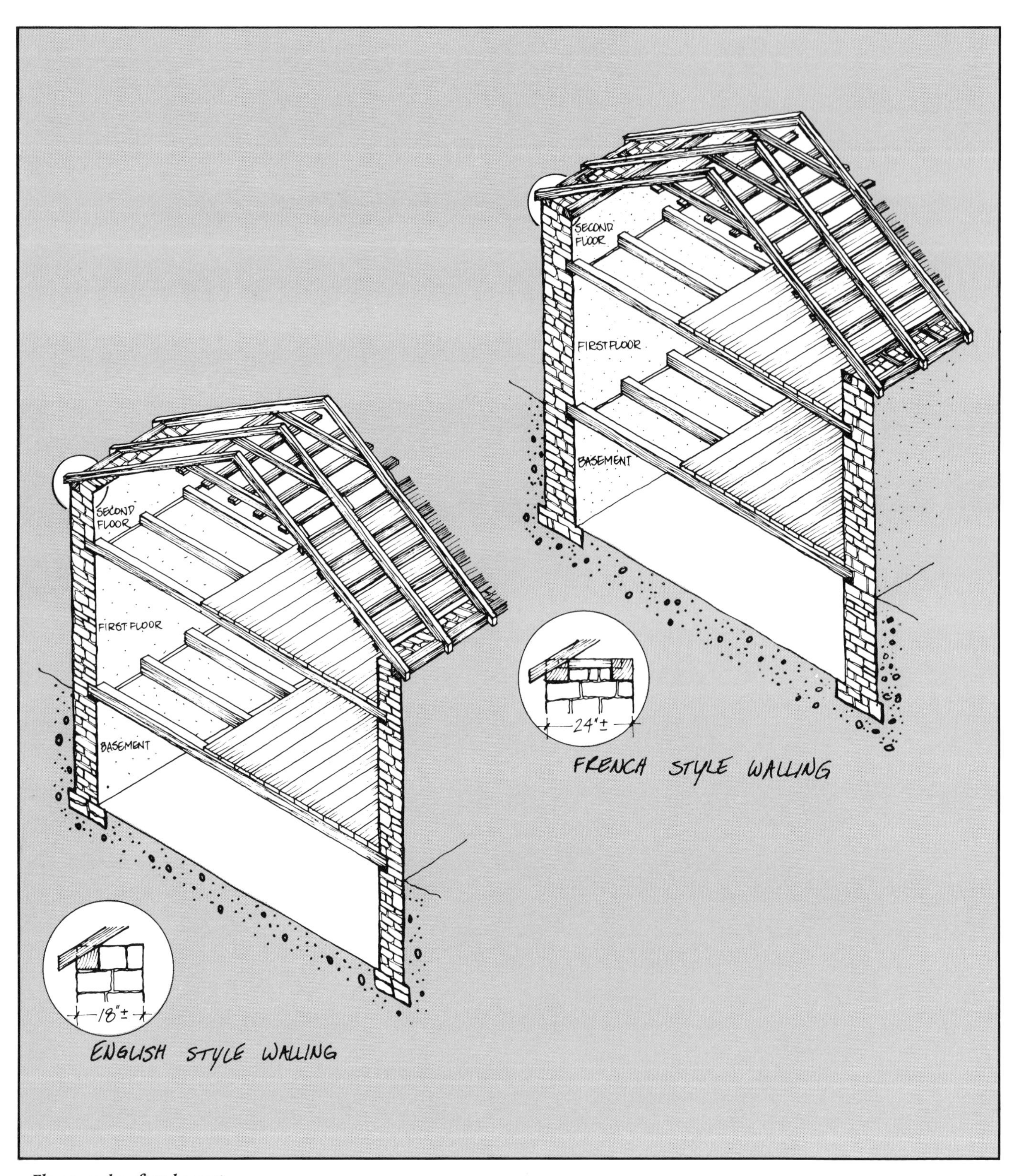

Floor and rafter layout

intend to build on bedrock or without a basement, be prepared to erect a rubblestone foundation of the proper dimensions. The ideal solution is to create a slipform style foundation by combining concrete and rubblestone. The trench should measure two feet wide and four-and-a-half feet deep, and the foundation should be designed to allow for water and sewage piping. Chapter 3 discusses insulation and drainage.

At this point, fir floor beams should be set into pockets in the wall. It is imperative to have at least one level area from which to take all measurements, a door sill on the main structure for example. A minor discrepancy now will become major as construction proceeds. Set up guidelines with mason's string to ensure that the stonework is erected in consistent, even planes.

Once the foundation is in place, chalk out the locations of doors and windows. Stone and mixing materials should be positioned at regular intervals around the site. Mortar should be mixed in the same proportions as in repointing (see page 70) but of a moister consistency. Large amounts of mortar will be needed, and renting a mixer is very worthwhile. Don't mix more than can be easily used in half an hour.

Building with stone should be done by at least two people, preferably working as a team. The actual construction process goes quite quickly, but the logistics of mixing mortar, carrying it to the scaffolding erected as the wall grows, selecting and bringing stone and all the other necessary small jobs can be quite time-consuming. It is important to define who will do what at the start of the project, because teamwork will ensure consistency in the finished walls.

Moving the stone onto the scaffold can be done with either a block and tackle or a hydraulic lift on the front of a tractor. Make sure your scaffold is not overloaded and that an adequate walkway is left open.

With this advice in mind, begin to put the first course into position. Wet down your first pile of stones and lay out a line of them parallel to the foundation. Their faces, though irregular in shape, should be flush in plane. The parts that sit in the mortar, called the bedding planes, should be adequately cushioned in the mortar. The stones should be sitting in a bed of mortar, not resting directly on top of the foundation. Trowel the mortar on two or three inches back from the face of the foundation so the weight of the stones won't squeeze the mortar out onto the face of the foundation. All joints should be at least one-half inch back from the faces of the stones; pointing will fill the remainder. The stones should be as close together as possible but with mortar between. Repeat the process along the exterior until the first course is completed.

At the same time, a similar procedure should be taking place along the interior wall. The void or cavity between interior and exterior walls must be filled with irregular-shaped stones, rubble and mortar. This cavity mortar is not exactly soupy but it should be very wet, allowing it to be rodded down to fill all voids. Do not underestimate how much rubble can be used; you will need quite a lot. The average stones taken from a historic building will probably have a measurement of not less than ten inches from front to rear. Theoretically, therefore, there will be a four-inch cavity in a two-foot-thick wall. In practice, the stones are irregular in depth, and consequently the width of the cavity varies. Wet all stones before use.

The addition wall must be attached to the existing structure. Using a two-pound sledge and a cold chisel, open up a void in the original wall right back to rubble-filled cavity. The opening will be irregular as it follows the shape of the stones you have removed. Build a new corner where the new and old walls meet. The new wall will be interlocked, both structurally and visually with the old.

As the wall continues course upon course, continually check your guide strings to make sure that the planes are even and consistent. The novice should periodically go to the corner of the building and look along the wall. Bulges will be obvious from

ABOVE

An orderly layout of materials makes for efficient, safe construction.

RIGHT

The mason is filling the core of the rubblestone wall with a combination of mortar and rubble. Note the guide strings and threaded rod put in place to accept the ridge cap.

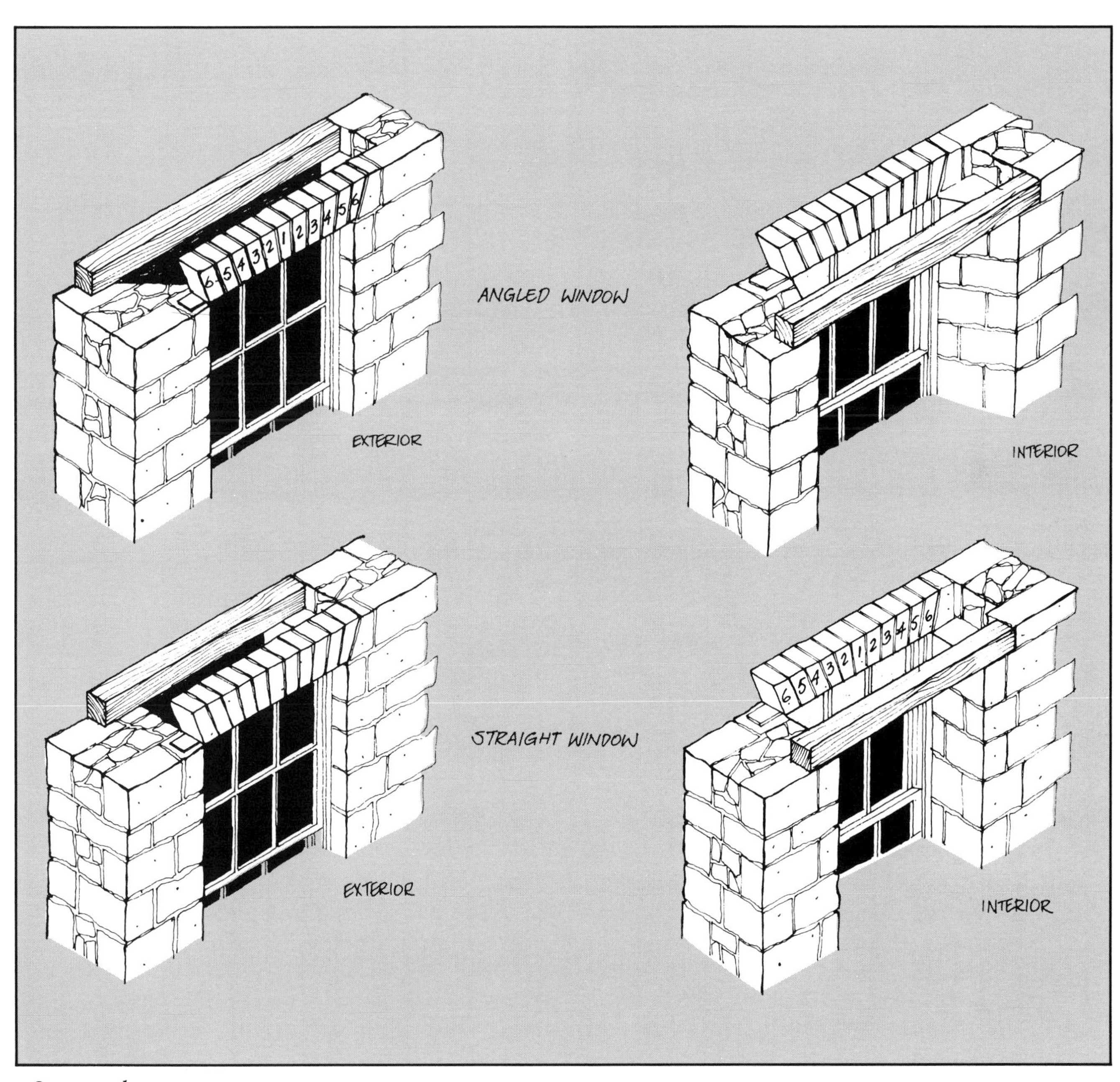

Stone arches

OPPOSITE

The wooden lintels are placed and a metal angle bracket is positioned to accept the stone arch. Note the guide lines.

Door and window details should be included according to traditional models.

A guide post set in each corner keeps horizontals and verticals absolute when building an addition. Note the wall plate (top right) on this uncoursed rubblestone wall.

this position. If you have built high enough, stand on top of the scaffolding and look down to cross-check. Also stand well away from the wall and take a long look at it. Working in close proximity for a period of time may make it difficult to see waves or irregularities unless you step back from time to time.

Every couple of courses introduce a bondstone that stretches the full width of interior and exterior walls; it links them into a solid mass. In a wall twenty-two feet long and fifteen feet high, every four courses, four feet apart, is a good guide. The stones in every course should cover the joints in the previous course, preventing continuous joints between courses. Until the mortar sets, there is relatively little cohesion strength in the wall. Resist the temptation to build too high in a single day, otherwise it will collapse; approximately five feet is the maximum. On the interior wall, depending on the type of finish to be used (see Chapter 6), mortar in wooden nailing strips one-half inch wide running horizontally every five feet. These are for attaching trim and interior cladding.

Historically, corners were often finished with dressed or at least rectangular stones called quoins. Their use is illustrated on page 25. Quoins are used only on the exterior, and although they are not absolutely necessary, their geometrical forms do ensure a certain degree of exactness.

Door and window sills can be made of dressed stone or wood. For the layman, wood is probably the easiest to work with, the most readily available and the cheapest. If you're building on a budget, door and window openings should be designed to accept standard units. Most old houses require custom sashes and doors, which can become quite expensive. It is often possible to buy old doors of the appropriate period and style for less than the cost of a new unit. Old windows, on the other hand, are difficult to find in sets, usually need extensive repairs and will result in heat loss without proper storms. The point is to know what you have or can easily get before designing these openings.

The first opening to occur in stonework is the door. Form the opening three to four courses (about one foot) above grade. A rough frame called a buck should be built in duplicate; one for the interior and one for the exterior. The buck can be made from spruce or pine, treated with a weatherproof sealer and primed. The size should be large enough to accommodate the sill, the trim and the door itself. Leave enough room; shims can be used later when you are actually hanging the door. Set the buck in place and support it with struts. Position it one inch back from the face of the stone. Lag bolts or anchors are driven through the frame so that they project into the wall cavity. As the wall is built up around the buck, the lag bolts take the place of the struts and the frame is held in place.

Above the rough buck, a lintel will be needed. This is a structural element that crosses an opening to provide lateral stability for the materials above it. A segmented stone lintel is probably the most historically accurate and attractive. To begin the process set a $\frac{1}{4}''$ thick, 4″ wide piece of primed steel on top of the frame but a half-inch back from the exterior wall. The steel should be long enough to extend four inches into the wall on either side of the opening. On top of this an arch is formed with a keystone as the middle element. Before the arch is positioned, a bed of mortar should be laid down, and the wall on both sides of the arch should be built up one stone back from the edge of the arch (see illustration). On the interior, two four-by-four pieces of ash or hemlock are installed in the same manner as the metal lintel.

Select the stones and lay them out on the scaffold, along with a small prybar, some wooden shims, a hatchet and a nimble-fingered assistant. It is a good practice to number the stones with chalk before building: keystone #1, stones on the left and right of it #2, stones on the left and right of that #3, etc. Start at the edges and work toward the middle. Once all the stones are in place, use the shims to position and angle them correctly. Place loose stones at the back of the arch, so that the face of the lintel is on the plane of the surrounding wall. Mortar the arch together; protruding shims can be trimmed off just behind the faces of the stones once the mortar has set. Pointing will fill the gaps.

The same procedure can be repeated for windows, exactly in the case of rectangular openings and with slight variation for an angled window well (see illustration). In the latter case, the wooden lintels are set into place at the rear, and the interior wall has a wider opening than the exterior. The use of metal in the lintel is a modern-day innovation. Once the arch has been correctly built and the mortar set, it becomes like a plug in a hole; it supports itself as well as the wall above it. Variations on arches and lintels can be found in the photographs illustrating Chapter 1.

At this point, begin to build the fireplace or chimney flue according to the instructions given in Chapter 5. Assuming that the addition is to have one floor only, finish the side walls and cap them on the exterior with a wall plate to accept the rafters. This plate is lag bolted into the wall cavity. The interior wall should carry on to become flush with the beam. Continue the gable wall in an ever-decreasing size until the desired pitch is achieved.

Joists should be set in place either by erecting a tripod and using a block and tackle or by hiring a crane for a morning. Make sure the walls are sufficiently dried to accept their weight. Roof sheathing, shingling, fenestration and interior finish can then proceed as in any type of construction. Pointing, as in Chapter 3, should wait until the walls are completely built.

The Brick Addition

As in building a stone addition, any brick structure added to a historic building should be in harmony with it, not jarring, not merely ordinary. The addition plan must accommodate your requirements, but the finished structure must be compatible in proportions, materials and fenestration. Building with brick takes planning and a certain amount of patience, but the three walls needed for an addition can be erected by the dedicated amateur.

The three basic traditional forms of brick wall are solid, veneer and cavity. Because the addition is being built today, the cavity type has been chosen. This method allows for contemporary insulation and systems. A second but equally important consideration is that it will allow you to match the bonding pattern of the original house.

For similar reasons, the foundation type chosen employs contemporary materials and methods. A fourteen-inch footing of reinforced concrete should be poured four feet below grade and a ten-inch block wall constructed on it. For additional strength, the cavities in the blocks may be poured solid with concrete. This is a particularly good idea if you decide to excavate for a full basement. Parging, waterproofing, insulation and drainage can be handled according to the directions in Chapter 3, although the parging material should be Sealbond mixed with ordinary Portland cement, rather than the traditional mortar mix (see Chapter 9 for details). For any exposed parging, however, use white Portland and Sealbond for a traditional appearance.

Mark out on the original rear wall where the addition wall will meet it. Remove every other original brick to the height of the new wall according to the illustration. When building the new wall, the face bricks may be interlocked with those of the original house. This method insures overall harmony in both line and bonding with the main building.

Now frame a two-by-four wall on what will be the floor of the interior portion of the addition, taking into account all windows and doors. Lay up the interior brick wall around the framework; it will act as a guide to keep the wall true. Attach rigid foam insulation sheets to the outside of the brickwork using adhesive. Make sure an air vent has been installed every eight feet before laying up the exterior brickwork.

On the ten-inch block wall, four inches have been taken up by interior brick and two inches by insulation. The remaining four inches accepts the exterior brickwork. The external brickwork is laid up as in the illustration. Troweling methods are similar to those discussed under the stone addition (page 89). Wet each brick in a bucket first, then lay the mortar on one-half inch back from the face. There should be mortar between the ends of the bricks as well, but don't try to bring the mortar right out to the face, pointing will take care of the gaps. Follow the illustrations for the exact method. Building mortar should follow the same traditional recipe given in Chapter 3, although it should be fairly moist when used for pointing.

Roof joists and sheathing may now be installed. This gives the structure rigidity and prevents undue lateral movement. Once the walls have had time to set, pointing may be done to match the original house.

OPPOSITE

Inadequate foundations lead to major structural problems. This coursed rubblestone summer kitchen is literally falling off the main house.

Although the illustrations that accompany this section give step-by-step instructions on how to build a brick addition, it is advisable to describe the layout or placement of bricks into a bonding pattern. Unlike rubblestone, where your "eye" or judgment is all that is needed as a guide for unit placement, brickwork deals with absolutes. In brick it is either level and true or totally unacceptable. The only area of leeway is in the amount of mortar placed between individual bricks. When approaching this problem, it is very worthwhile to construct a small test wall. It should tell you very quickly how to achieve a joint compatible with the original building.

Working out a bonding pattern for the addition I have described can be simply done by devising a scale drawing showing every brick in each of the three walls. Window, door, stringer and corner details should all be carefully included. The drawing will give you the feel of building and identify potential problem areas beforehand. Brick cannot be laid out one wall at a time, because the interrelationship between the walls is what holds them together. Your three addition walls must be interlocked at the corners. This is unlike traditional rubblestone and, in fact, more similar to building with wood. As in the test wall to determine joint thickness, it is a good idea to practise building a lintel and a corner before running into problems halfway up the actual addition wall. The brick lintel is similar in construction to stone (see page 93), but very complex examples may require the use of plywood forms to support the bricks until the mortar has set.

Make any changes to your drawing that seem necessary, then transfer the grid to the foam insulation on the wall using chalk or pencil. You don't have to draw in every brick, but block out areas and their brick types. For example, red bricks may form the body of the walls, while buff bricks or yellow marls might be used in corner, door and window details. This will tell you how these areas relate to the rest of the structure.

Attention to period plaster, woodwork, lighting and furnishings succeed in making this addition compatible with the main house.

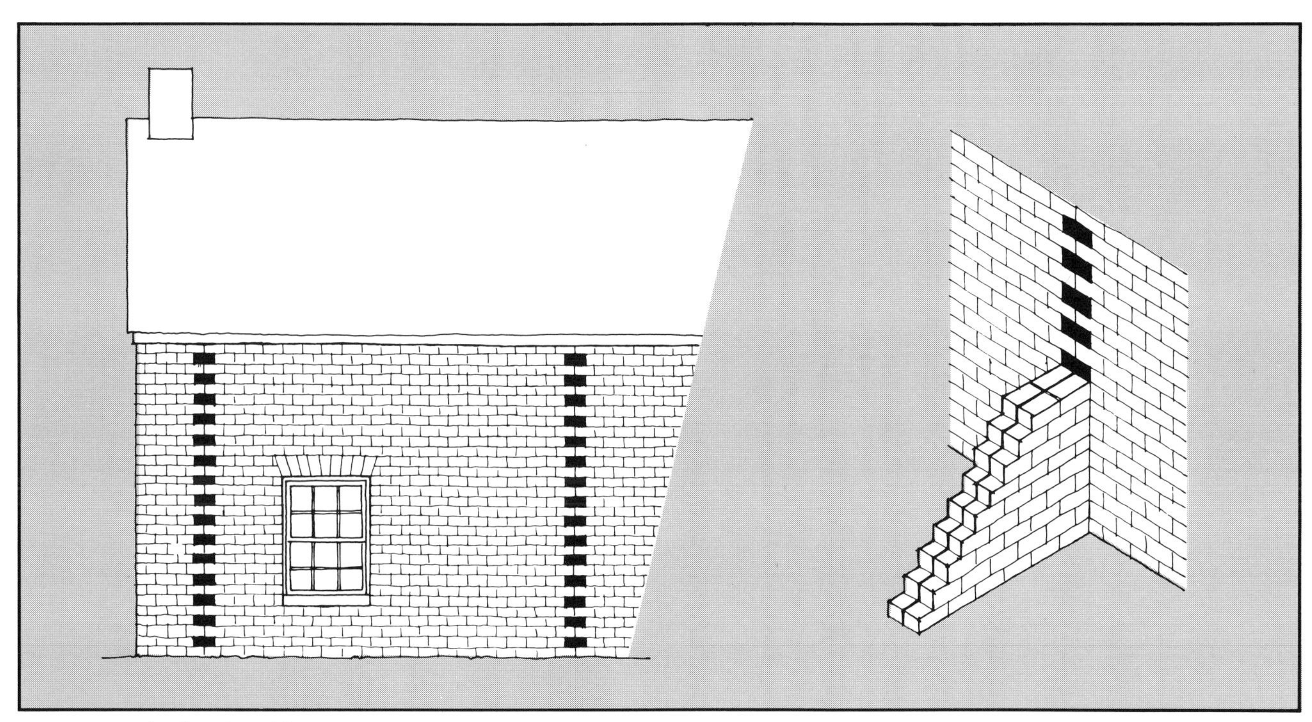

Tying in the brick addition

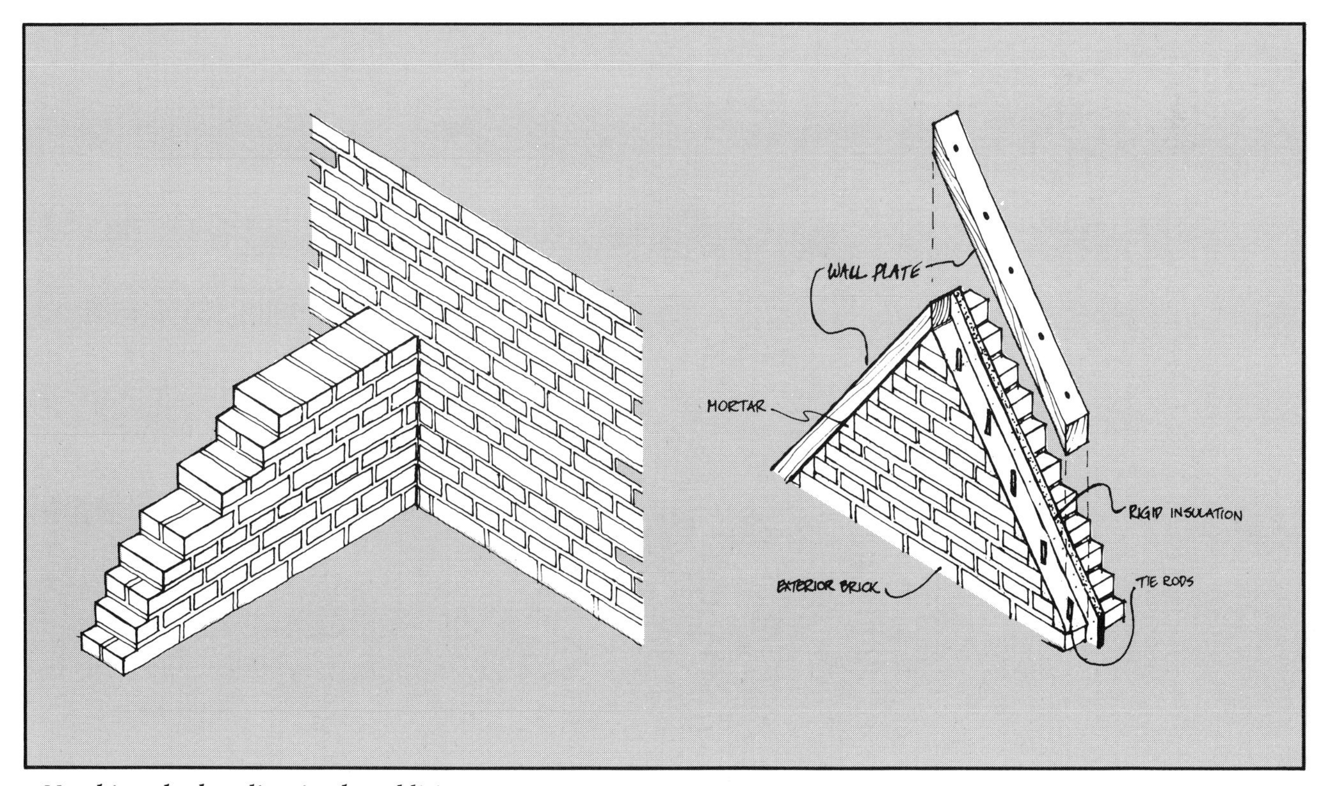

Matching the bonding in the addition

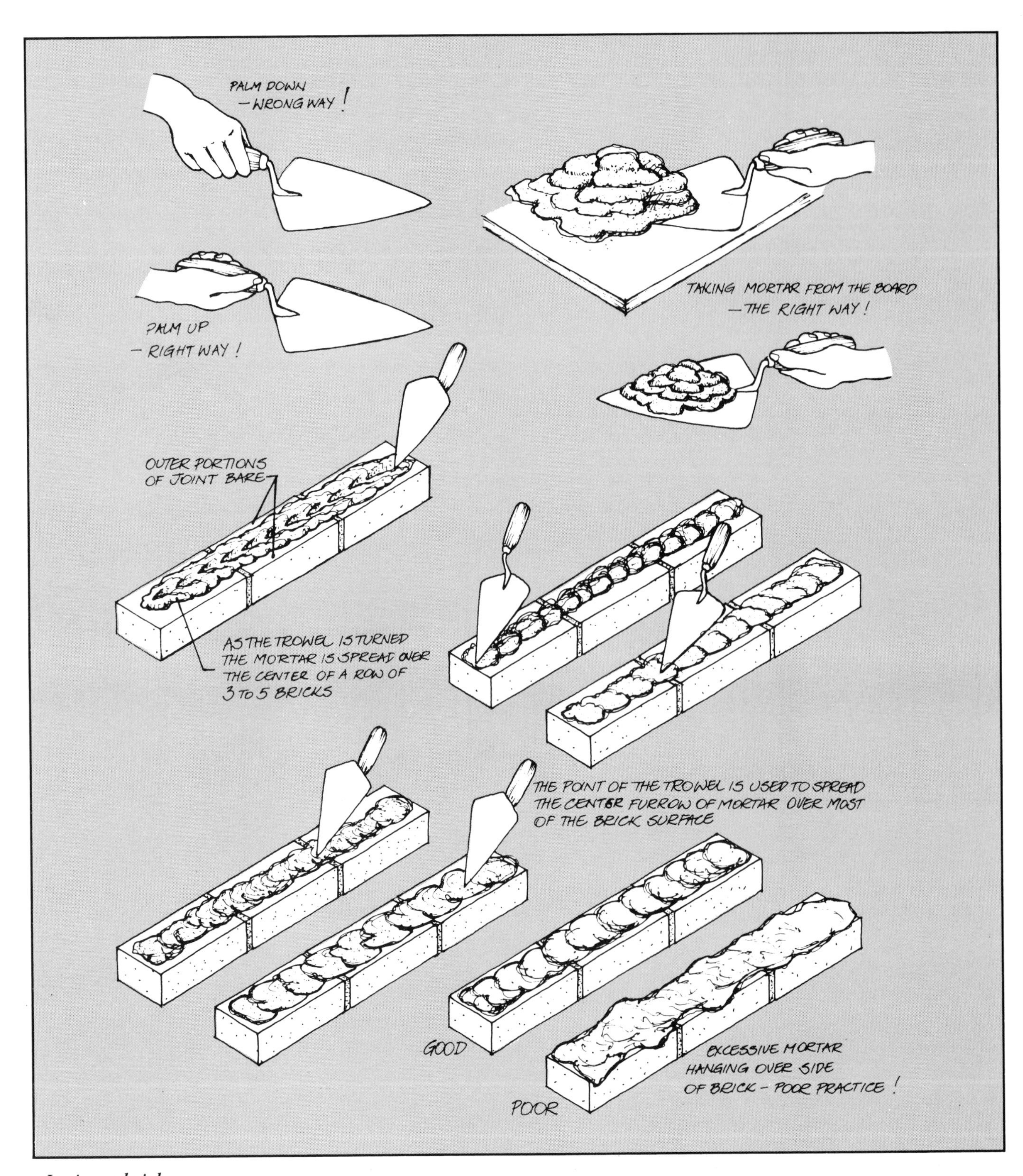

Laying a brick

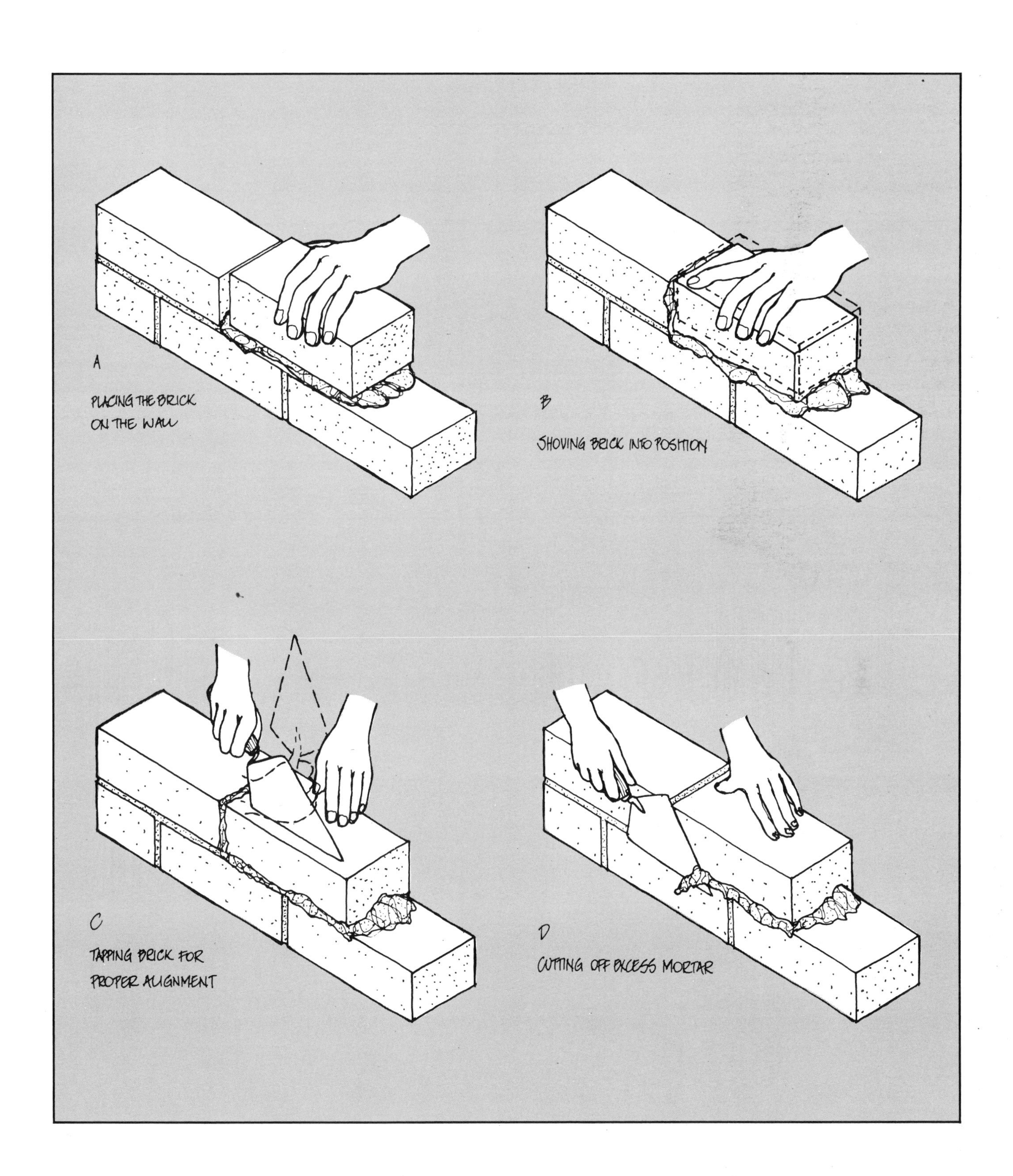
A
PLACING THE BRICK
ON THE WALL
B
SHOVING BRICK INTO POSITION
C
TAPPING BRICK FOR
PROPER ALIGNMENT
D
CUTTING OFF EXCESS MORTAR

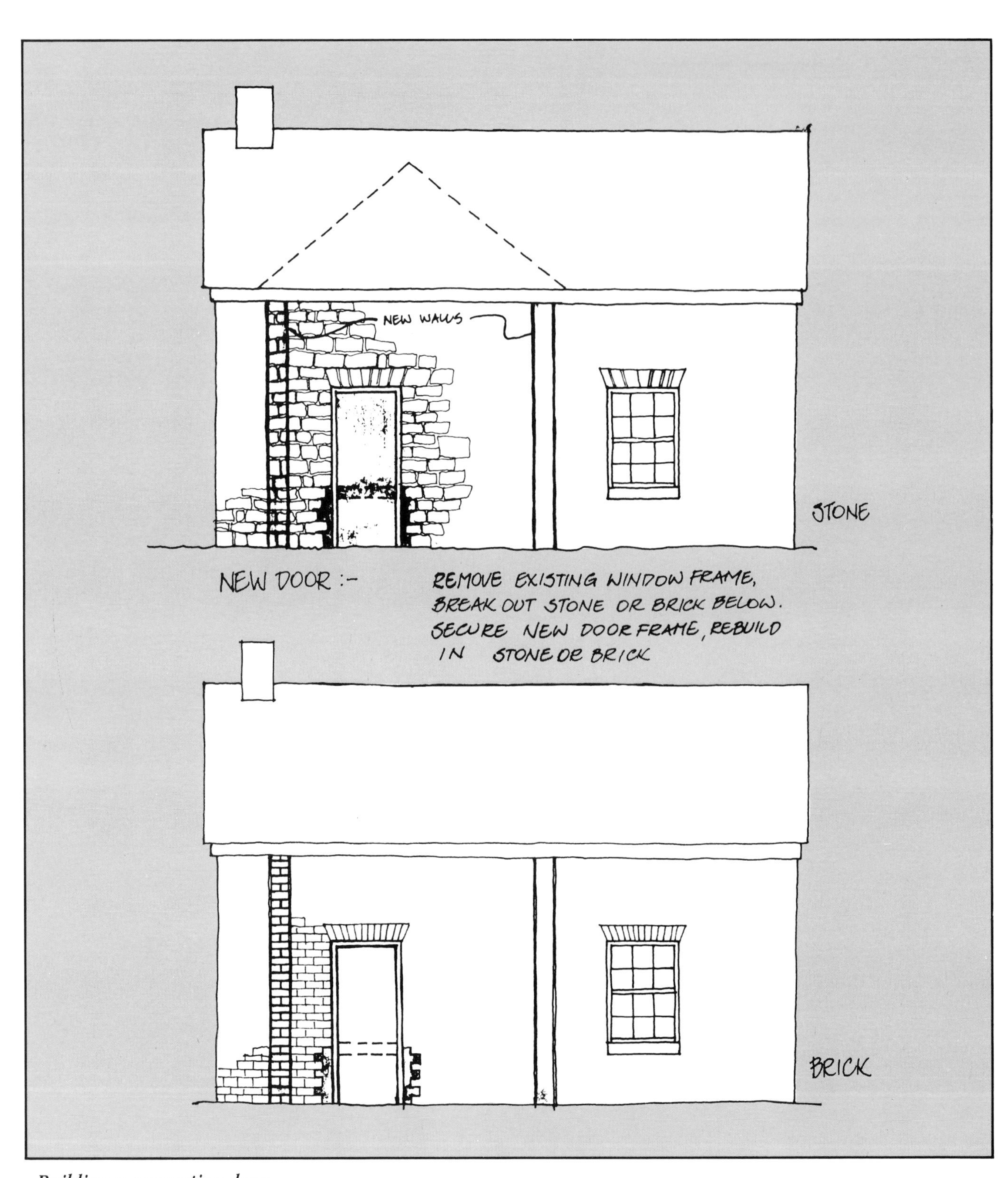

Building a connecting door

A contemporary addition to a mid-nineteenth-century house.

If the layout appears satisfactory, construction can begin in earnest. Make sure the drawings don't weather off the insulation. Keep your drawing handy. Tape it to a piece of plywood and cover it with plastic. That way it won't end up as a revolting mess and will be suitable for framing when you have finished. Take your time, enjoy the process and keep harmony with the original in mind.

The Connecting Door

When a stone or brick addition is attached to the rear wall of a house, the conversion of an original window into a connecting door is a very practical solution to an otherwise complex problem. The width of the window usually approximates the width of a small door, and the lintel above already supports the masonry above the opening.

Carefully remove the window sash. It can be recycled in the addition. Take out the rough buck and the wooden or stone sill. If the sill is stone, chisel out the mortar around it and work one end toward the interior of the house with a prybar. The center of the sill will act as a pivot point. Stone sills are heavy, and this is a two-person job. When it is free and balanced in the opening, it can be easily moved.

Remove the masonry under the window opening in much the same way that the new and old walls were joined. Install a rough frame for the new door; it will act as a form for rebuilding masonry edges.

CHAPTER FIVE

Chimneys, Fireplaces and Stoves

The Neoclassical form at its finest with a semi-elliptical cut stone arch, four end chimneys and overall harmony of proportion. Prescott, Ontario, first quarter, nineteenth century.

SOME OF THE MOST JOYFUL elements of the period house are the heating and cooking systems employed throughout its history. The romance of the fireplace is always a central focus in any restoration, while the friendly warmth of the cast-iron stove has regained its popularity for obvious economic and environmental reasons. Most preservationists who are involved in their own houses are interested in resurrecting original heating methods as back-up or secondary systems. In today's energy-conscious society, this is not only romantic but good common sense.

However, systems used in the past must be thoroughly understood before they are implemented. Their history, function, construction and maintenance are all important aspects that must be mastered. Unlike modern heating systems, old systems can be hazardous. But they are also a major factor in re-creating the mood of the past. New heating systems can be included in the period masonry building without dramatically affecting its original shape, as long as they are compatible with historical forms and technically efficient.

As in many other parts of this book, to begin we should start on the roof. A simple smoke hole was the forerunner of the chimney; it not only acted as a vent, but also as a type of "window" to admit light. By the Middle Ages these openings were called louvers, in part at least because of their illumination qualities. The smoke, stench and dirt were so offensive that by 1526 the scullions of Henry VIII were ordered not to work in the kitchens "naked or in garments of such vileness as they now do,"[1] despite the fetid atmosphere.

In the normal European domestic structure, the fire was usually set on stone in the center of the house, and the smoke allowed to drift through any available opening. The introduction of the vertical gable wall with a built-in fireplace must have been a very welcome innovation. Wall fireplaces, in fact, had first appeared in the late eleventh century,[2] but little technical evolution took place until the late 1700s. Wattle and daub (basically sticks and mud) were the major chimney materials, despite the obvious fire hazard.

Wattle and daub chimneys were very popular for several centuries. Throughout the early settlements of New England and Canada, the method was employed until the mid-nineteenth century. Well into the twentieth century in Appalachia, it was common practice to build a firebox of fieldstone and tie it into a flue of four-inch-wide wooden slats chinked with clay.

The ever-present threat of fire quickly made brick or stone a necessity in chimney and flue construction in areas where masons and materials were available. As early as 1737, specifications for work on two houses in Louisbourg, Cape Breton Island, required that

> all the chimneys will be built of good brick, eight *pouces* long by four wide, cleanly laid, and fully bonded with a mortar of lime and fine sand. Their flues will be plastered with the same mortar as neatly as possible.[3]

Specifications in flue construction, as well as regulations governing the use of wooden shingles, were doubly reinforced by a 1721 Montreal fire in which 138 buildings were either damaged or completely destroyed. It is difficult for us to grasp today how ever-present the threat was, and how much it was feared. The parapet walls often used as an architectural feature of free-standing or rural buildings were originally developed as firewalls in early communities.

Chimney flues, or funnels as they were first called, were minus a damper until the changes introduced by Count Rumford and Benjamin Franklin. Until 1750, ovens were often built into the backs of cooking fireplaces, and as a result cranes for hanging pots were not employed because they impeded bake-oven operations. Instead, implements were hung from lug poles or trammel chains that could be adjusted away

An eighteenth-century Quebec cooking fireplace with a bake oven set into the rear wall.

from the swing of the bake-oven door. The bake oven later moved to a position beside the actual firebox, which by the mid to late eighteenth century had become raked or angled in the Rumford manner. This allowed cranes to be used which pivoted on lugs mortared into the firebox. It, and the steaming pots it held, could be swung away from the fire. Some bake ovens protruded through the rear wall of the chimney, forming a beehive shape on an exterior wall.

As the nineteenth century brought technological change, cooking and heating were relegated to the cast-iron woodstove. The original flue system might be used, but the inefficient and labor-intensive fireplace was closed off. The "modern" introduction of the stove was greeted with mixed feelings, and its removal from the domestic scene was vividly described in the 1851 comments of one Arthur Channing Downs, Jr. Writing in *The Horticulturist*, Downs observed that:

> [The use of the stove] is a growing evil, far more serious in the Eastern States than we in the West can well imagine. In my visit to an Eastern city, the loss of the open fires is everywhere oppressively felt — furnaces, furnaces, nothing but furnaces — no bright, cheerful fires to enliven the scene — all dull and gloomy, exhausted and exhausting.... What are to become of [hearths] and their genial associations of social ties and social joys: are they all to be swept away? When far from home, where does fancy picture dear ones? — surely around the blazing fire. When memory calls up scenes of early childhood, are they not of the same place, whence we looked up into the faces of dear parents? Yes, all the recollections of boyhood and manhood are all connected most pleasantly together at this spot, and the hearthstone becomes sacred to us all — we love it, we cherish it, and, if needs be, we would fight for it.[4]

The fireplace, the flue and the chimney certainly do form the core of the house, both in sentimental and historically utilitarian terms. Probably no other mechanical element of the house is subject to the same extremes of heat and cold. Lack of maintenance or physical abuse make it imperative to inspect the

A Northwest cookstove similar to those manufactured at the Percival Stove and Steam Pipe Company, Merrickville, Canada West (Ontario), in about 1840. The small heater at the left is for heating flat irons.

A Neoclassical parlor stove.

OPPOSITE, ABOVE

A cooking fireplace with appropriate utensils and implements. The brickwork is a running bond. Eastfield Village, New York, late eighteenth century.

OPPOSITE, BELOW

A cast-iron Franklin inset placed in a larger fireplace opening. Eastfield Village, New York, 1780.

entire system before initial use, followed by a yearly maintenance routine.

Chimneys and their pots must be examined for cracked or loose mortar. Wherever possible the flues themselves should be inspected for structural cracks, a procedure that may be impossible because most are behind walls. The flues may be plugged, either from early chimneys collapsing from within, or the fact that they were never cleaned during the entire history of the house. Parging or mortar joints may be worn or loose, and fireboxes may require repointing or rebuilding.

A good initial test, although a messy one, is to build a small fire with softwood kindling. Once the fire catches, blanket but don't smother it with wet leaves or some other moist, combustible material that will smoke. The smoke should be drawn easily up the flue. If not, the blockage cannot be far away. At this stage, place a damp blanket over the chimney opening. If there are any leaks, smoke will pour out of every void in the flue system.

Perhaps the best way to illustrate the importance of fireplace inspection, and the frustrations it may cause, is through a personal anecdote. Early in my experience as an old house enthusiast, I acquired a delightful rubblestone house with barn attached. There were only two rooms downstairs and a large sleeping area above. On the ground floor, covered by paneling and subsequent plaster, was an eight-foot-wide cooking fireplace and adjoining bake oven. Both had obviously been hidden early in the history of the house.

Needless to say, I was delighted with my discovery. As I removed the rubble from firebox and oven, I discovered that the flue was absolutely solid with stone, ashes and birds' nests. The astonishing thing was that the house had been heated with a spaceheater connected to this chimney. Somehow the flue had accepted the waste and the heat without poisoning anyone or starting a fire.

Over a three-month period I cleaned the flue completely, ending in a mad fit on the roof where I drove a TV aerial up and down in the flue to dislodge the final two feet of debris. The final bit collapsed, a minor explosion of dust poured from windows and doors, but the flue was clear at last. The next two days were spent clearing up the mess in great anticipation of the first fire.

That first fire was set, very justifiably in my own mind, using lath and debris from the fireplace. With a great sucking sound, the heat was drawn up the chimney. It worked! As I sat in front of it, gleefully throwing on more fuel, I happened to glance out the window. Smoke was everywhere. I ran outside and saw to my dismay that smoke was pouring from every joint in the exterior end wall, culminating in great billows around the soffit and fascia. Inside and upstairs the smoke was doing the same. As quickly as possible I extinguished the fire and considered the lesson. Miraculously, the house did not ignite. Every joint in the building was worn out and needed repointing, let alone the parging necessary on the interior of the chimney.

The remedial procedures I followed haven't changed very much over the intervening and much safer years. The hearth and firebox should be repointed using the appropriate historic mortar mix (see pages 64 to 73). Building codes will call for the use of modern firebrick in the firebox, but you may be allowed to meet standards by overcompensating with historical materials. For example, eight inches of soft clay brick may be viewed as the same as four inches of firebrick. Check the code and talk to your building inspector.

The highest concentration of heat in the fireplace occurs on the slope of the rear wall, and this is usually the area that most needs repointing and general repair. In stone fireplaces that are badly worn, remove the offending stonework gently with a small prybar and two-pound hammer. You may have to

A finely cut, early nineteenth-century stone chimney and cap with narrow masonry joints.

shatter one stone so the others can be removed easily. Rebuild the firebox following the methods described in Chapter 4. Especially important to remember is that the stones should form a smooth, even plane for the smoke to flow over; the mortar joints should be flush and the stone faces smooth. Brick fireboxes, more common because of the clean faces of the bricks, can be repaired in much the same manner. But when repointing be careful not to mar the soft edges or dislodge surrounding brickwork. Tooling of the joints should be flush, but adhere to the methods used in the remainder of the fireplace. The back walls of most brick or stone fireboxes were originally constructed on a slope or arc, and some forming may be necessary in rebuilding. Mortar stains can be removed with muriatic acid. Cut the acid with water according to the directions on the bottle, apply it with an old brush, let it foam for a few minutes, then wash it off. Vent the room and wear goggles; the fumes are poisonous and caustic.

If your fireplace has constant or near constant use, a free-standing metal fireback is advisable. Antique firebacks cast specifically for the purpose are rare, although reproductions are made by companies specializing in cast iron.[5] Historically, stove plates were commonly used for the same purpose, and these are easier to find. No matter what the source, a fireback has a certain historical charm, but it also reflects more heat into the room and protects the brick or stone in the back wall.

Most early fireplaces were not built with dampers. While you are rebuilding the firebox, it would be worthwhile to expose the area directly above to install both a damper and a smoke shelf. The damper will prevent excessive heat loss when the fireplace is not in use and allow you to regulate the draft to some degree. A smoke shelf (see illustration) will cause a slight downdraft, which will draw the smoke up the chimney more effectively.

Fireboards were sometimes used to cover the fireplace opening, acting as a type of damper. In a contemporary context, the back of a fireboard can be insulated with rigid polystyrene. When combined with a damper, an insulated fireboard can be very effective against heat loss, but do not insulate if any other flues run into the chimney. Do not replace a fireboard until the fire in the hearth is completely out.

The actual installation of a damper and smoke shelf in brick or stone fireplaces is illustrated. Approximately one foot above the metal lintel, remove a section of wall large enough to let you work in the flue. The metal damper should be put in at the one-foot point. Its shape and positioning should follow the illustration closely. The flue size should not be reduced in any way, and the mechanism of the new damper should be as simple as possible to prevent clogging and jamming. Many late Victorian dampers used a crank and cog mechanism. In my experience, most of these do not work effectively because of lack of maintenance, and they are costly to repair.

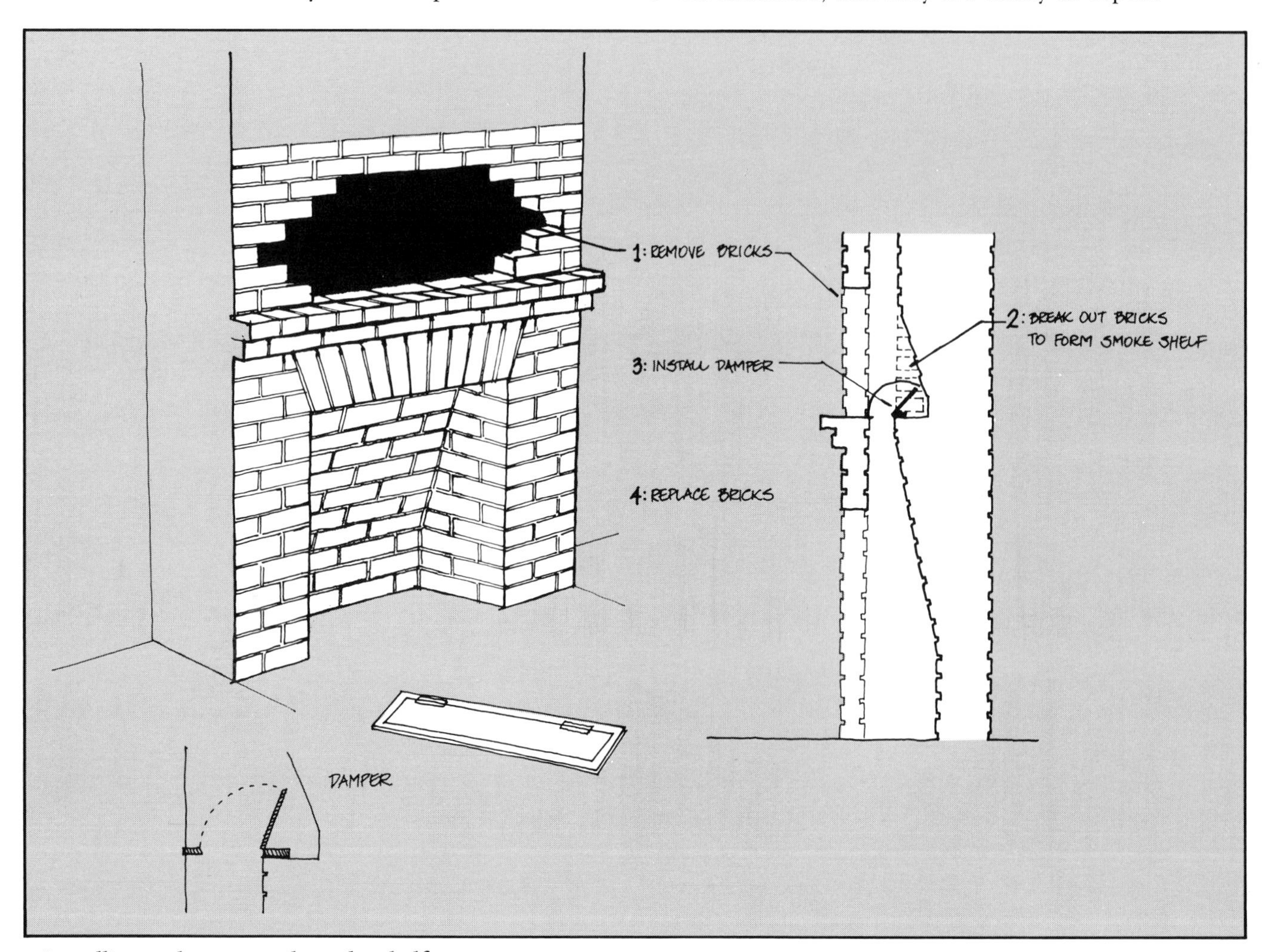

Installing a damper and smoke shelf

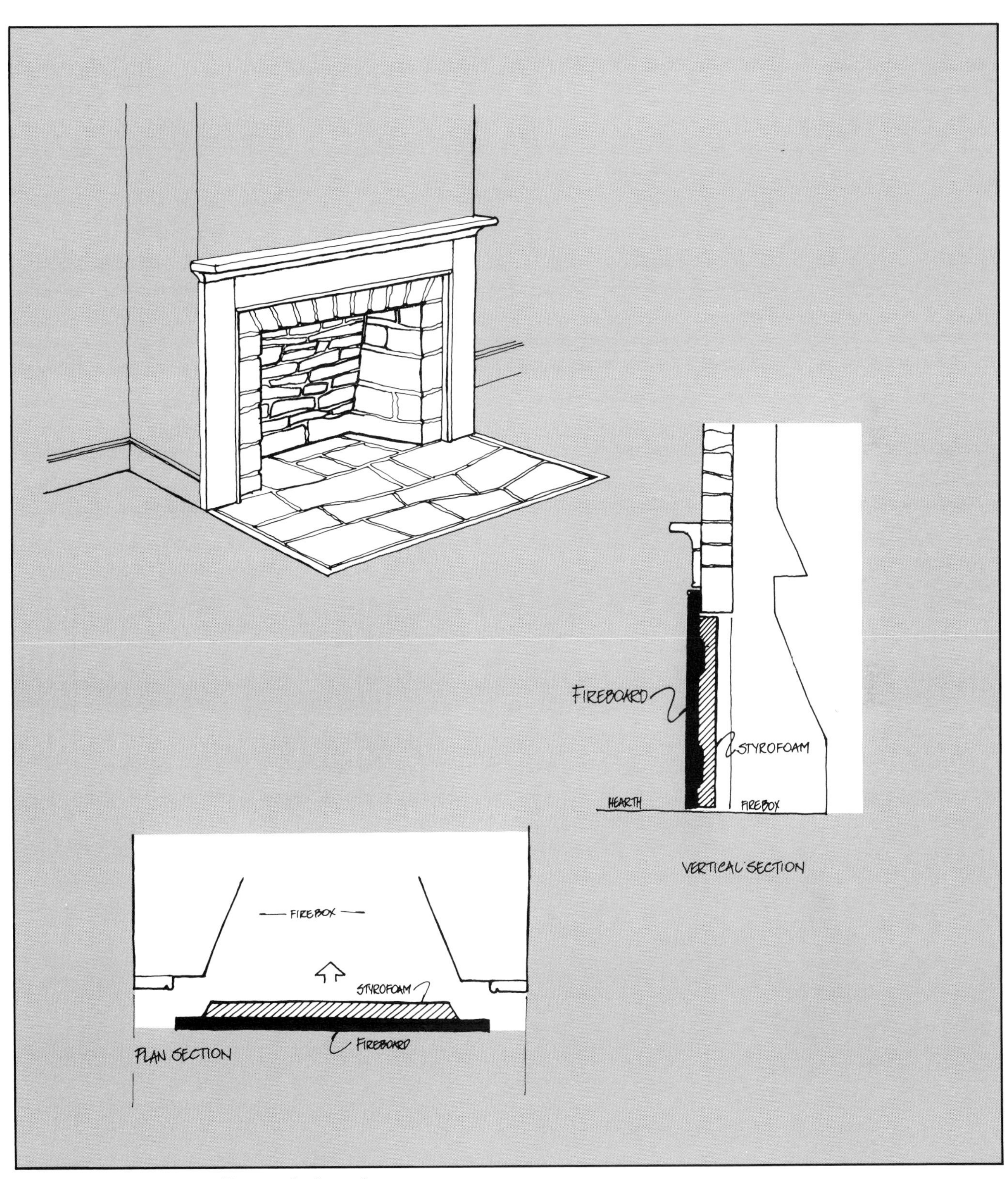

Installing an energy-efficient fireboard

Create the smoke shelf in the same plane as the damper. It should be at least four inches wide and cut into the rear wall of the smoke dome or flue. When working on brick, be careful not to disturb the overall wall unit. Most brick fireplaces were constructed in a running bond style. Stone fireplaces are invariably thicker and consequently less delicate.

When the job is finished, remove all debris and parge around the area on which you have worked. If the flue above the damper has already been cleaned by a professional and appears to function properly after close inspection, yearly maintenance with a wire brush will usually prevent problems. If, on the other hand, the fireplace has not been used for many years, major repointing and parging will probably be necessary. Before either of these operations can take place, the brick flue must be cleaned with a wire brush or by sandblasting on stone. Creosote clogs joints and surfaces making it difficult for new materials to adhere.

Most chimneys and flues were built on interior walls, and therefore they must be excavated from within. As in installing a damper, a hole must be opened into the flue. The bonding methods used in the original brick or stone must be respected when doing this so rebuilding is a straightforward process. Repoint and reparge where necessary using the mortar mix described on page 164. Be very careful not to reduce the overall flue size, although minor variations will occur if you decide to install modern flue tiles instead of traditional parging. The following method has proven effective when using these contemporary units. Open the entire face of the chimney. Flue tiles come in standard sizes: 8″ × 12″, 12″ × 12″ and 16″ × 16″. Place two tiles of the appropriate dimensions side by side on a small metal T just above the smoke shelf, mortar them into position, and continue the procedure to the roof. Finally, rebuild the face of the chimney.

Working on flues is a dirty, awkward business at the best of times. Some imaginative, though not terribly safe methods have been devised to skirt the inherent problems. On one particularly large, particularly early and particularly worn-out example, I once saw a slim lad lowered down the flue on a rope. This prevented breaking holes in the wall, was vigorously applauded by all but the boy involved, and did work. However effective it might have been, it was totally unsafe and is not recommended!

Assuming that you have used accepted methods, once the flue has been adequately treated the wall will have to be rebuilt with traditional materials and methods. Bear in mind that you will have to parge the inside as you build. Remember, too, that the average temperature in a domestic flue for a fireplace or airtight woodstove is 600 degrees Fahrenheit (315°C). Do a good job.

Where the flue protrudes from the roof is the next area of concern. Adequate flashing must be installed as in the illustrations. When hot air travels up the flue it cools, when it hits the cold air outside it condenses on the chimney pots. Therefore, a stronger mortar is advised in chimney construction (see Chapter 9). Where chimney caps were originally employed but have since deteriorated, new ones made of reinforced concrete should be installed.

Until the early nineteenth century, the fireplace in North America was the primary source of heat and the only element for cooking. The early, large cooking fireplace, smoky and grossly inefficient as a heat source, would often have two or three separate fires built within it for the cooking of different dishes. Although Benjamin Franklin's new designs of the mid-1700s, and Count Rumford's innovations shortly thereafter, no doubt improved the cook's lot, the invention of the stove for cooking and heating must have been a godsend.

Heating stoves began to be used in North America by the late 1700s, and by 1820 they were relatively common in established areas. Early stoves were built in pieces. The base with legs attached had a lip around it which accepted the four side plates. These, in turn,

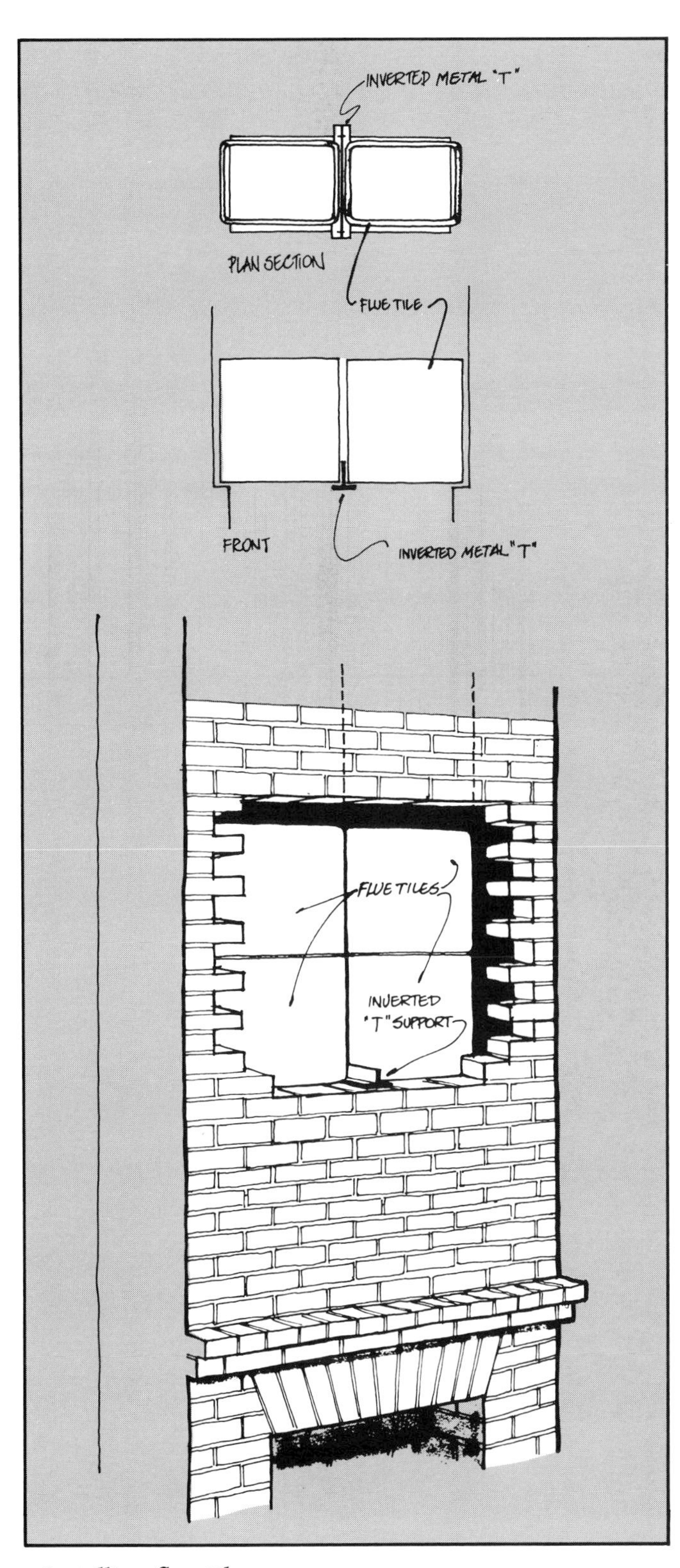

Installing flue tiles

slotted into each other. The top held the entire assembly together. The primary reason behind this type of construction was the problem of transporting heavy, bulky items from manufacturer to consumer, although relatively easy disassembly for cleaning was a side benefit. Compared to today's airtight stoves, they consumed large amounts of fuel due to the drafts around every joint. Despite this problem, early stoves were far ahead of the old fireplaces.

Stoves were set around the house — parlor, dining room and so on — frequently standing on brick hearths or iron floor protectors. Often, the stove was set into the old fireplace and its flue directed into the existing opening. Ductwork ran throughout the house, and in some areas heat holes were cut to allow warm air to pass into other rooms. Stove design paralleled architectural fashion, and the Classical Revival of the nineteenth century spawned its share of models with the columns and pediments of Greece and Rome.

Stoves made specifically for cooking were common by about 1830, and they proliferated as the century progressed. The average price in the Upper Canada of the '30s was £15 to £20, very substantial amounts for the time. The double stove was a notably Canadian variation, which was reserved only for the rich, or at least the well-heeled. It was a single, large unit made up of two separate stoves: the lower one was an expanded unit designed to accept larger than normal pieces of wood, while the upper portion served as an oven.

The intense heat generated by the kitchen stove was no doubt welcome in winter, but summer must have been a different matter altogether. The sweltering cooks of early Victorian North America were a direct cause behind the development of the architectural addition known as the summer kitchen.

Rochester, New York, was famous for the fine stoves made there, and they were widely distributed throughout the northeastern United States and Canada. The Industrial Revolution, however, quickly led to wide-

The roofless building in the foreground was the separate kitchen. This arrangement was the exception rather than the rule throughout the northeastern United States and Canada by the mid-eighteenth-century.

spread foundry activity and the subsequent localization of stovemaking. More mechanized techniques and local manufacturers meant prices went down and stoves could be tailored to specific areas. The Percival Stove & Plough Company of Merrickville, Ontario, which is still in business as The Alloy Foundry, made a two-tiered cooking stove in the North West style for eager consumers on the fringes of settlement. J.R. Armstrong & Co., King Street, Toronto, listed their Bang-Up stove, "better adapted to the town where saving of fuel is an object — the oven being large and high." Perhaps the last word on living with these period devices is contained in this delightful if somewhat chauvinistic comment contained in a book by author Jeanne Minhinnick: "The grandest sight in the world is the woman you love taking an apple pie from the oven."[6]

Due to their relative inefficiency, you may not choose to install period cast-iron stoves in your masonry home. Although nineteenth-century stoves did and will consume larger amounts of wood than twentieth-century airtight models, they are cleaner burning. That is, an airtight stove will produce significantly greater deposits of potentially combustible creosote, particularly if it is routinely operated with the damper turned down. All stove piping should be regularly and religiously cleaned, but pay special attention to pipes attached to the airtight variety. My own practice with our "modern" stoves is to run them full open with a good fire at least once a day in the winter. This is not a substitute for cleaning, but it will retard creosote buildup. Alternate energy is efficient — make it safe.

As fossil fuels and electricity became the norm for heating, many early masonry buildings suffered the loss of fireplaces or stove chimneys. If you own such a house and are thinking of returning it to its former glory, it is important to build with traditional models in mind. The remainder of this chapter is devoted to methods of new construction with period charm.

In the first example, we will be building a Rumford-style fireplace. If the original fireplace has been removed at some point in time, the brick or rubblestone base will probably still be intact beneath the floor. Such a survival is fortunate, because it will give you a good indication of the original dimensions. After checking the base for structural damage and load-bearing ability, you should be able to build upon it.

If an entirely new base is to be constructed, choose your location carefully. Whenever possible, build on an inside wall. Although it is less convenient, building on the inside is preferable, because heat retained in the masonry mass will probably lessen fuel bills. Heat the inside, not the outside. Before you start to

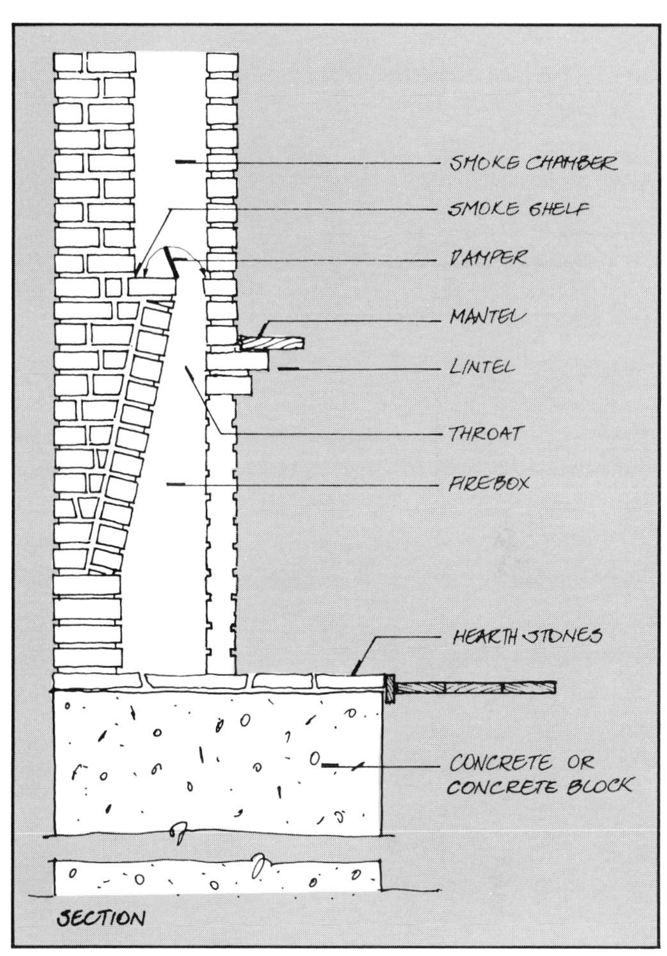

Rumford fireplace

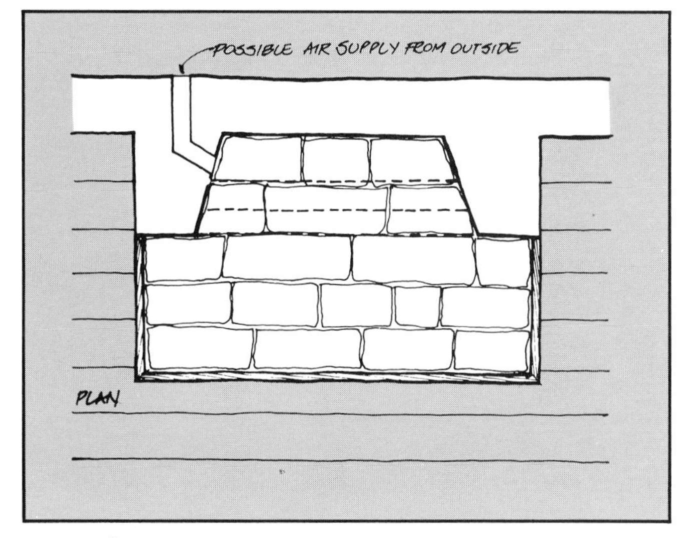

Hearth pattern

RIGHT

When the foundation and hearth have been set in place, the side and back of the firebox can start to be formed. Note the tooled hearth stone.

BELOW, LEFT

The walls and back follow the guide lines that have been positioned to show the absolute dimensions necessary when building a Rumford fireplace.

BELOW, RIGHT

The metal lintel is set in place, the stone arch is built on it and the damper unit is set into place above that.

OPPOSITE

A finished early nineteenth-century stone fireplace with wooden mantel and surround.

build, check all the way through the building (floors, ceilings, rafters, etc.) to make absolutely sure that satisfactory clearances are possible, and adequate structural support is available as the chimney rises through to the roof.

As with any masonry mass, the weight of hearth, firebox and flue must be adequately supported from below. A base fifty-six inches wide and seventy-six inches long would be needed to bear the weight of a fireplace with an opening four feet by four feet and a hearth extending twenty inches from the face of the opening into the floor. The base will have to be built on a footing at least six inches larger in length and width than the base made of concrete block poured solid (all holes filled with cement) or reinforced concrete. The new base should end approximately six inches below the finished floor. This allows the hearthstones or bricks to be set into place flush with the floor in a historically correct manner.

If you are building with stone, lay down a thick bed of lime-based mortar (see page 164) and place the stones in it, the fewer the better (consult illustration). Brick hearth patterns vary widely; check the neighborhood for survivals in houses of the same period, take a few photographs and imitate one that appeals to you. Once the hearth has had time to set, chalk out the profile of the finished fireplace as in the illustration. The firebox and flue, whether of brick or stone, should be constructed according to Count Rumford's principles:

- The width of the firebox must equal the depth.
- The vertical portion of the firebox must equal the width.
- The thickness of the firebox must be at least two and a quarter inches. (I use at least twelve inches of brick or stone.)
- The area of the firebox opening must not exceed ten times the flue opening area.
- Both the width of the entire fireplace and its height should be two to three times the depth of the firebox.
- The opening height should not be larger than the width.
- The throat should be no less than three or more than four inches wide.
- The centerline of the throat must align with the centerline of the firebox base.
- The smoke shelf should be four inches wide.
- The width of the lintel should be no less than four and no more than five inches.
- The vertical distance from lintel to throat must be at least twelve inches.
- A flat plate damper is required at the throat and must open toward the smoke shelf.

Follow the illustrations and captions to finish the job up to the roof.

Many chimney styles have been used in the past. Again, it is probably best to check around the immediate vicinity for surviving examples. Pay close attention to capping styles on stone and corbelling methods on brick. Corbelling is a masons' technique to enlarge a structure or opening by stepping successive courses. Particularly in the mid to late Victorian era, some corbelling became very fancy, almost whimsical, employing several colors and bonding patterns.

In our second example we will be building a relatively simple flue and chimney that is ideal for use with a woodstove. Before beginning, though, check the house thoroughly to make sure no surviving flues have been covered over. One house, built about 1865, originally had twin chimneys for several woodstoves. The chimney sections were knocked off in the 1930s, and the roof was patched. The new owner found that the original suspended-bracket flues were still intact in the attic. With a bit of repointing, they became quite usable. The entire job only required new external chimneys. This owner was lucky. Most bracket

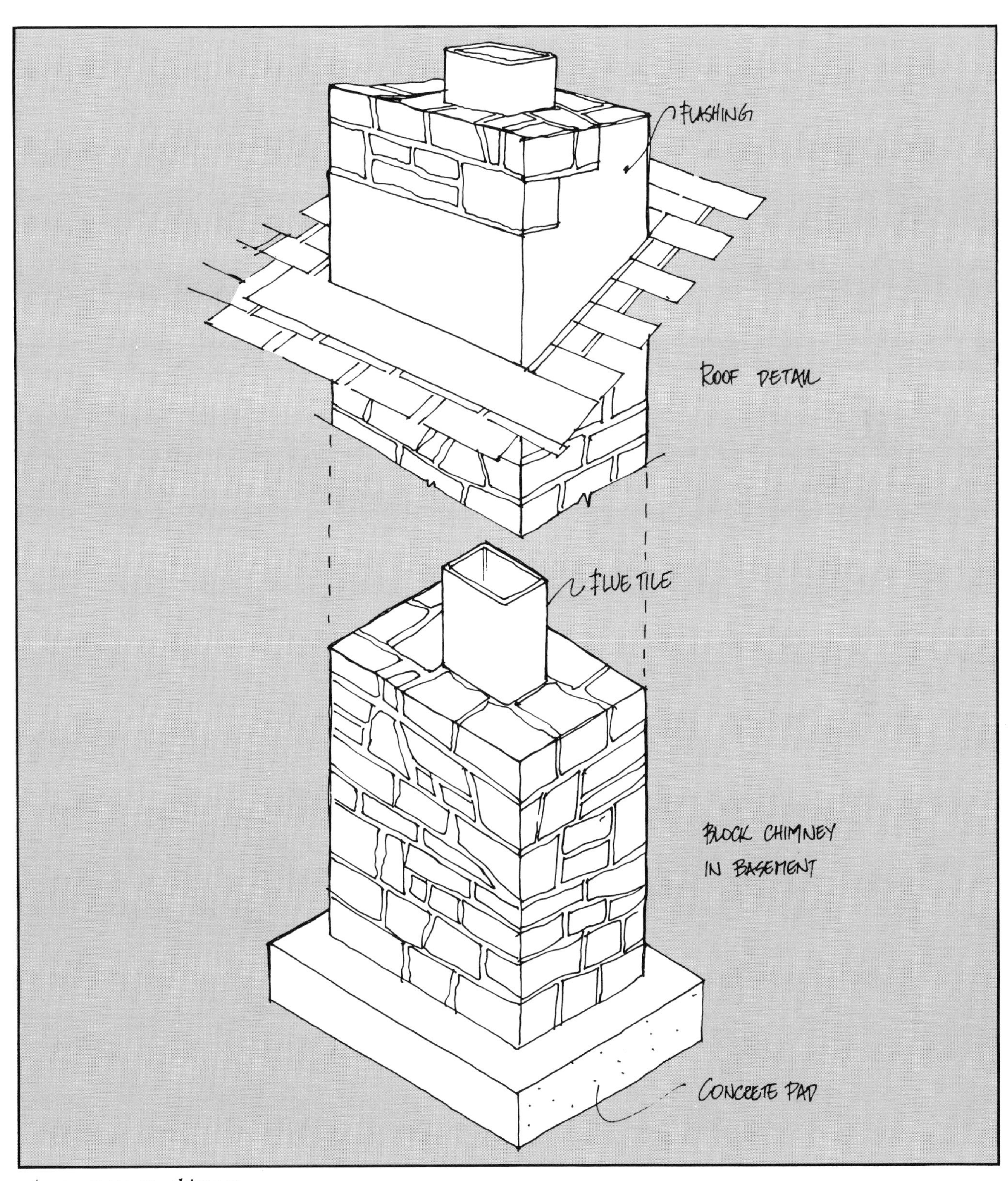

A new masonry chimney

chimneys are supported on wood with less than a two-inch gap between masonry and combustible material. This is light years away from contemporary building codes, and replacement is necessary. As in our first example, build the new flue and chimney by following the illustrations and notes.

Whether you are repairing an existing period heating system or building a new one along traditional lines, it is important to acquire any hardware beforehand. Test it thoroughly to make sure it is the right size and of approximately the correct date. Be sure of your measurements. Ordering a damper will do no good at all if it turns out to be too large or too small. Antique cranes are very interesting additions to a fireplace, but they do require that pivot lugs be mortared into the firebox. The effect of a beautiful crane will be quickly spoiled if it does not have sufficient radius to swing without banging into the firebox wall. Bake-oven doors, antique or reproduction, should be made of metal and have a metal frame.

A cobblestone house.

CHAPTER SIX

Interior Finishes

Decor and architectural style complement each other in this mid-nineteenth-century dining area. Note the plaster medallion on the ceiling.

THE INTERIOR OF THE historic masonry house is the one area that most non-professional preservationists feel is their forte. Elements of decor are often well within their scope, and appropriate interior treatments will be a daily reminder of their talents and a daily reinforcement of their understanding of the past. The interior is made up of three major elements: floors, general joinery or woodwork; ceilings and walls. More often than not, the last of these involves the use of plasterwork, a craft directly related to that of the mason.

Although there is a very real temptation to begin working on the interior as soon as possible, it should be staunchly resisted until you are certain that the exterior is in the best possible condition. You will be making more work for yourself in the long run if that elegantly plastered wall is destroyed by moisture penetration from poor external masonry joints.

A balloon or timber-frame building supports the roof and floor by some method of tying the elements together, as in mortise and tenon joints. Most masonry support systems, on the other hand, rely on pockets and built-in sills to accept joists, beams, rafters and so on. The major difference between the interiors of wooden-frame and masonry houses is wall thickness and mass. A frame building uses wooden uprights and parallels to create a web for attaching interior cladding and finishes. The wall profile of a nineteenth-century timber-frame building might be made up of exterior cladding, perhaps clapboard, nailed to plank sheathing, which is attached directly to the framework. On the interior, lathing might be nailed onto the other side of the framework, with plaster applied as a finish.

Two types of interior construction are found in masonry houses. In the first, furring strips or nailing blocks, called dooks, were inset into the solid stone or brick wall to accept baseboards, chair rails or picture rails. Interior joinery, including window and door trim, was applied directly to the interior surface. The rest of the framing, lathwork and joinery was constructed and applied to the partition walls. When the time came for finishing, plaster was applied directly to the outside walls without the use of lath. The relatively rough masonry surface provided an adequate base to which the plaster could key. Lath was nailed to framed-in interior walls, and all plaster was butted to the previously installed joinery.

In the second type of interior treatment, battens were applied to the solid masonry walls, lath was nailed to the battens and plaster was spread over the lath, butting to the joinery. The obvious advantage for the modern preservationist is that there is a slight airspace between the wall and the lath which provides room for electrical wires, plumbing and, in some cases, insulation.

Both of these methods have plaster as a finish in common. The use of plaster in clay, lime and gypsum form has been recorded from early times, but the most familiar type in North American period houses is that applied to hand-split lath, sometimes called accordion, or the machine-made variety.

Many interior plasters were made up of three coats, although there is ample evidence that two coats was the norm in some areas. Plasterwork generally suffers from lack of maintenance and old age, as does the rest of the structure. Cracks, hollow-sounding bulges or damp stains should be investigated immediately. They may be due to structural movement or

OPPOSITE

An early nineteenth-century bedroom. The stone chimney has been whitewashed to blend in with the rest of the room.

OVERLEAF, LEFT

A restored mid-nineteenth-century bedroom featuring basic plasterwork accentuated by window trim, a chair rail and a baseboard.

OVERLEAF, RIGHT

An irresistible dinner invitation in this converted loft space.

ABOVE

This coursed rubblestone partition wall shows accordion lath and plaster on the left and on the right door details and nailing strips called dooks.

LEFT

Plaster was applied after the joinery. In the case of this mid-nineteenth-century example, the plasterwork attached to machine-cut lath stops at the baseboard.

PREVIOUS PAGE

The Van Alen House in the Hudson Valley of New York State, built in the early eighteenth century, is one of the purest surviving examples of Continental Dutch architecture in North America. N. HUTCHINS

BELOW

A Prince Edward Island coursed rubblestone, one-and-a-half story house in the Scottish tradition, circa 1840. SCOTT SMITH

This mid-nineteenth-century Italianate brick house employs both the color scheme and architectural elements made popular by A.J. Downing. N. HUTCHINS

A fine late eighteenth-century Georgian home built with solid brick walls. N. HUTCHINS

The break-up of mass in this Georgian house is accomplished by the use of a pronounced cut-stone water table above the foundation and parapet fencing on the roof. N. HUTCHINS

OPPOSITE, ABOVE

An early dry sink cut into a piece of limestone. PHILIP JAGO

OPPOSITE, BELOW

The furnishings in this restored 1820 stone house have been carefully assembled to convey a period mood. ART HOLBROOK

ABOVE

The Dutch in New York's Hudson Valley often used brick and rubblestone in combination. N. HUTCHINS

BELOW

This new addition employs materials and building techniques of the mid-nineteenth century. JAMES T. WILLS

This 1840 house, Neoclassical in form, uses simple yet highly decorative bonding styles and lintel treatments constructed from locally made brick. N. HUTCHINS

OPPOSITE

A rubblestone wall with vertically positioned rubble capping. N. HUTCHINS

Space and light are used to full effect in this urban loft conversion. BRIAN GLUCKSTEIN, DESIGNER, CRESFORD DEVELOPMENTS; PHOTO BY D. BURKE

climatic change causing broken plaster keys. The original plaster may have been incorrectly applied, for example, if the lath was nailed too close together so that the plaster could not squeeze between individual pieces to form proper keys. Early repairs may have used improper materials. Excessive moisture, either from leakage into the interior or inadequate vapor barriers, may encourage fungus growth. Salts present in exterior walls may transfer to the plaster through capillary action; the result is general weakening and staining.

When repairing plasterwork, it is imperative that all areas be totally free of moisture. However, a problem often arises when a house has been heated in the past with a less than effective system. A new owner installs a much more efficient system with the intention of completing the plastering, painting and so on during the winter. The original plaster has contained a certain amount of moisture throughout the history of the building. The new heating unit quickly bakes out all the moisture, with cracking and general failure the inevitable result. Take warning and dry your house out slowly.

Plaster repairs on a small scale, or even large projects, can be readily undertaken by the layman. Lime, sand and hair binder were the major materials used in plastering throughout domestic North American architecture, although gypsum plaster did become popular during the early nineteenth century. Gypsum was not a new development; it had been mined around Paris during the Middle Ages, and, probably as a result, the term "plaster of Paris" was coined.

As in the restoration of all period elements, care must be taken in plastering to employ materials, tools and methods as much in keeping with the original as possible. But before looking at repairs, recipes, and full plastering procedures, it would be worthwhile to discuss the condition of the interior in general.

Plastering should be one of the last things done. It comes after the support systems have been made sound, after trim has been applied, but before wall or floor decoration. Plank floors were always meant to have some give in them. Excessive bounce, on the other hand, may be caused by rotted joist ends where they are inset into masonry walls. Check and repair them where necessary. Sagging floors may be corrected to some degree by the installation of steel jack posts, so long as any raising or leveling is done slowly over a period of a week or more.

Period joinery may have begun to lose its detailing because of many, many layers of paint. If you want to return to the crisp outline of the original and have a good base for new paint, stripping is sometimes the only answer. About the fastest and cleanest method is to use a heat gun (see Chapter 9). Most oil-based paints applied in the past will contain lead, so vent the room completely and do not work for long periods of time. One benefit of working down to a firm painting base is that it should be possible to discern the original colors used on the trim. Stripping to the "natural wood" is not recommended as part of this procedure, because unpainted joinery simply was not a part of period decor. Similarly, softwood floors in eighteenth- and nineteenth-century North American homes were painted, not sanded and varnished.

Once you are satisfied that no extensive structural repairs or cleaning will be necessary on the interior, plaster repair or installation can proceed. The size and shape of accordion lath may vary widely, but the standard size of machine-made lath is $\frac{5}{16}''$ by $1\frac{1}{2}''$ by $4'$. These strips were usually nailed one-quarter of an inch apart, the gap being wide enough for the plaster to be forced between them to form a key on the inside. Basic plaster repairs can be done effectively with durabond, a commercially available joint filler. Cut the crack back to sound mortar and wash it out using an old window-cleaner bottle filled with water. Mix the filler for small cracks on a wooden palette and force it in with a putty knife. Larger cracks may require two quick-setting base coats. Cover the second one with drywall tape and go over it with a much thinned coat of filler. A plasterer's rectangular laying-on trowel is probably best for this kind of repair, because its size will leave fewer tool marks than a smaller putty knife. Don't try to fill too large an area at one time or new cracking will occur as the durabond dries. A plasterer's technique for limiting the amount of dusty sanding necessary to remove bumps is to smooth the repair with a wet rag before it dries completely.

Bulges can be successfully reattached to the lath using screws and large washers, which are countersunk into the plaster as the screws bite into the lath. If a large area is repaired using this method, a grid should be drawn locating the lath. The bulge should be forced back into position with wooden braces and the screws attached where indicated on the grid. Taping, as in repairing cracks, should be done and allowed to set up. When dry, the bracing should be removed, and a finish skim coat of durabond applied to the repaired area.

When electrical services are introduced into the period house, wiring often needs to be run through the walls. Holes are cut for switches and outlets. Once the plaster and lath have been cut, there is no backing to accept new plasterwork. The simplest repair method is to remove a complete strip of lath to the next upright stud on an interior wall. Cut a piece of metal lath (a cheap, contemporary alternative) in a strip slightly larger than the exposed opening. Place it inside the wall so the mesh overlaps the existing wooden lath, and toenail it to the stud next to the switch or outlet and the stud on the far side of the opening. A first coat of durabond will key the metal to the wooden lath. After setting up, a finish coat can be applied.

There is another method for repairing this problem, and those of a larger nature where a substantial portion of plaster has fallen from the wall. In the case of the smaller void, reattach the lath to provide a base, and then in either case install a piece of drywall over the old lath and up to the thickness of the old

ABOVE

New joinery is installed before the plasterwork is applied in this twentieth-century addition to a mid-nineteenth-century house.

OPPOSITE

The semi-circular arch and inset panel joinery frames the entrance to the music room in this mid-nineteenth-century house.

plaster. Fill the joints, tape them and apply a thin finish coat over the entire area.

Whenever original wooden lath is to be reused, it must be properly prepared before new materials are applied. If it is left *in situ*, be sure to wet it thoroughly. If it has been removed, soak it and clean it with a wire brush before renailing. The plaster will have clogged the pores of the wood, and the lath will be less than receptive to new materials unless these procedures are followed.

The modern-day preservationist has a major series of problems to examine when dealing with interior finishes that have been applied directly to masonry. As may well be imagined, dampness in these walls is a prevalent problem. Mildew and stains may discolor paint or wallpaper. In fact, wallpaper binders can fail completely in this situation. Contemporary mechanical systems like ductwork, wiring, plumbing and insulation can be installed in one of two ways. In the first method, mechanical systems must be relegated to partition walls only; insulation and vapor barriers on external walls must be ignored. If decorative or historical features are part of these walls, it may be best to leave them alone in any case.

In the second procedure, all joinery must be removed from the external walls, and new interior walls framed from within. These new walls will provide space for insulation, wiring and so on and allow the installation of adequate vapor barriers. This may be the lesser of two evils. Building a few new walls is probably preferable to disturbing many partition walls that are pure in their architectural form. The obvious disadvantage is that the overall size of any room treated in this way will be reduced, but the advantages may outweigh this one negative aspect.

When dealing with external masonry walls that have a gap between the lath and masonry, it is often possible to route mechanical systems with relatively little disturbance. By removing baseboards, a skillful electrician may be able to fish wires through these walls.

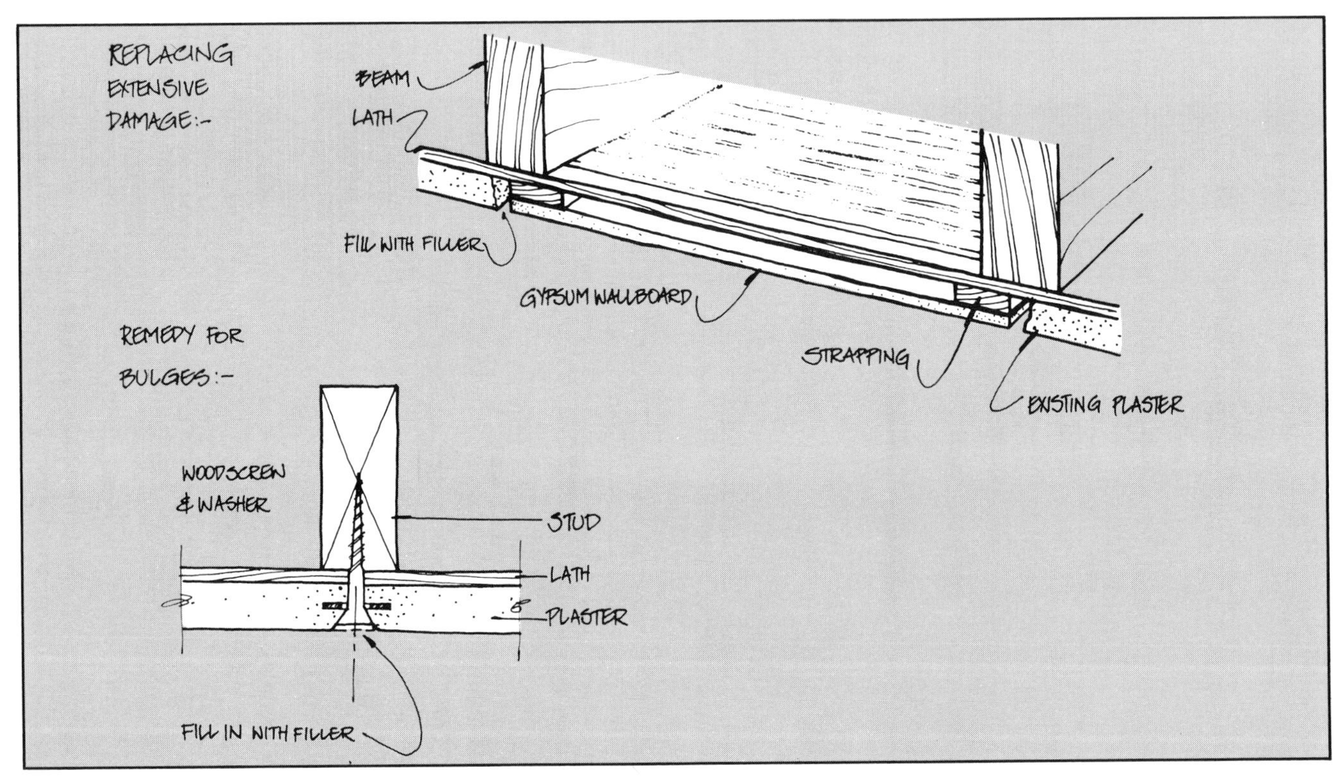

Plaster repair

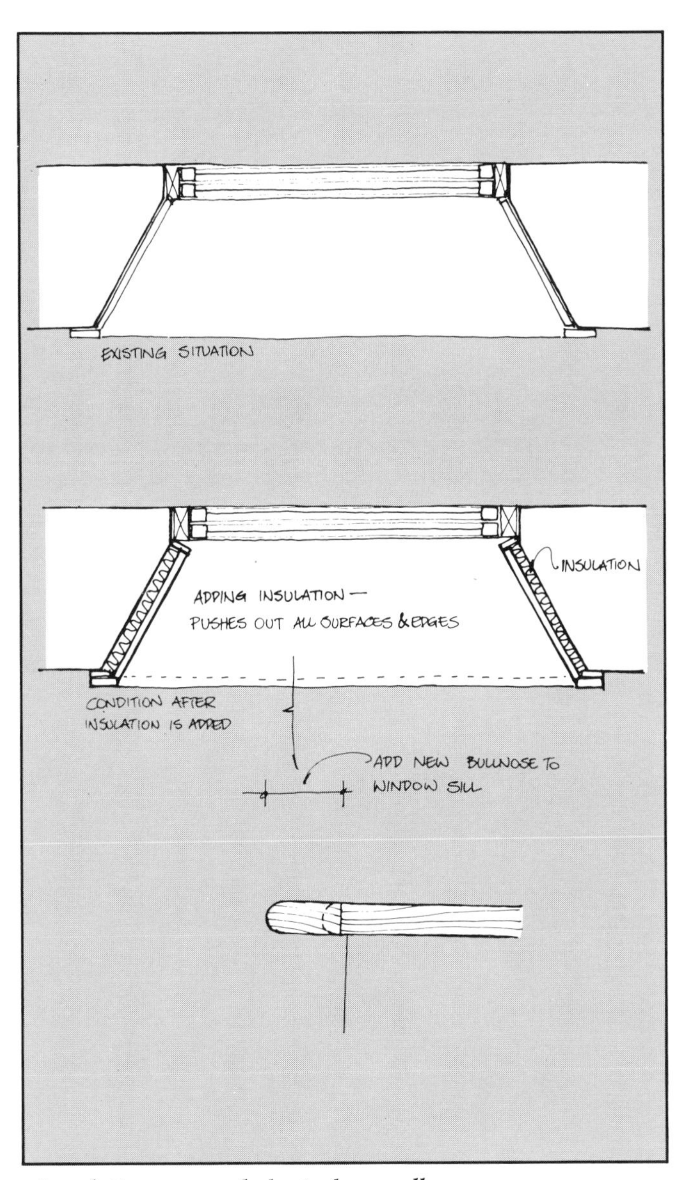

Insulating an angled window well

Although it is not advisable to run plumbing on exterior walls because of freezing problems, the space may accept normal bathroom piping. As in the first wall type, the less disturbance the better. In houses with good interior condition, they are probably best left alone when it comes to insulation. Where insulation is required, the most efficient, albeit the most expensive and extensive alteration of historical proportions, is stripping and reframing.

It is possible to lose as little as one inch when reframing with standard studs and finishing with three-coat plaster or half-inch drywall. Reframe as in new construction, allowing for all insulation and mechanical systems. This is an all-or-nothing proposition. Without impeccable vapor barriers, condensation will inevitably cause rot in the woodwork. Where breaks occur in the barrier, tape them to ensure the best possible seal.

One difficult area, both in architectural proportion and potential heat loss, is a room's windows. There is little insulation value in window returns, either of the angled or rectangular kind. The slightness of the cavity behind window paneling does not permit sufficient insulation, and when reframing there seems to be no solution short of redesigning the units. My own experience has shown that the best alternative when working on angled windows is to remove the original trim, frame around the opening and insert a compatible return between the original panel and the new interior wall (see illustration). As can be imagined, the dimensions of a splayed window will increase as it is extended into the room, and elements of original trim will have to be reproduced to fit the new size.

Sills are the most difficult to extend. The only method offering some historical integrity as well as practicality is to splice an extra piece to the original. Often, the sill is rounded, and a rectangular plane must be created to accept the splice. The splice, in turn, should be rounded to reproduce the original profile.

Insulate window wells by stuffing friction-fit batting between the paneling and the wall. This is not completely effective, but it does provide some degree of heat-loss protection. A metallic, aluminum paint may then be applied to the trim to form a vapor barrier. After struggling with the problem of retarding heat loss from the inside, you will begin to understand why areas around windows and doors have often been repointed on the outside.

The most important aspect of plastering, whether on completely new walls, repairs or new applications on an old base, is to duplicate the mood, texture and integrity of the original work. The approximate thickness of historical three-coat plaster is about one-half to three-quarters of an inch. The irregularity of the finish found when plaster has been applied directly to masonry walls often leads to the supposition that most interior finishes were irregular or sloppy. The majority that were left irregular were probably second coats, a third finish coat was probably planned but never executed.

The preservationist has a variety of choices when attempting to create the mood that only plaster can convey. For the purist, lath can be hand-split from pine or spruce. Contemporary wire lath is a viable and readily available alternative. River sand can be gathered, combined with hydrated lime and, during mixing, a hair binder can be added. Historically, the proportions would be three lime to one sand and a bit of animal hair gotten from the local tannery or *abattoir*. The set-up times are two weeks for the first coat, eight weeks for the second coat and eight weeks for the third. A more expedient plaster formula uses perlited gypsum plaster, sand for texture and a bit of animal hair. Set-up times are two days for the first coat, one day for the second and one day for the last.

An elegant contemporary bedroom in this converted space.

Three-coat interior plaster is applied in much the same way as the stucco described in Chapter 3. Slap on the scratch coat, which should be about one-quarter of an inch thick, using a hawk and a laying-on trowel. Score this coat to accept the second, or brown coat; it should be about one-eighth of an inch thick. The brown coat is not completely regular, but it should form a smooth plane. Once it has set up, a third or finish coat can be applied. Most novices tend to mix plaster too dry. It should stick to the surface but slip easily from the trowel. The plaster should be slightly wetter for the second and third coats. In this form the last coat is much easier to trowel smooth, although completely smooth plaster is an ideal that is never achieved. The final coat should be fairly thin, about one-eighth of an inch, but you'll find that keeping to any of these measurements over an entire wall is quite difficult. Plaster should cure thoroughly before painting. A month or more would not be unreasonable.

Yet another method works quite well in reframed rooms or on completely new construction in an addition. Cover the framework, insulation and vapor barrier with drywall and tape the seams. Any molding, wainscoting or chair rails should be nailed on at this point. Ready-mixed durabond that comes in tubs, not the dry variety, can now be applied using a fuzzy paint roller. It should be the consistency of thick soup; thin with water if necessary. Areas around joinery and the taped joints should be dealt with first, then ceiling and walls. Allow the material to stiffen slightly and trowel it as smooth as the inexperienced hand is capable of doing. You will, in effect, have created a slightly irregular, non-Spanish style stucco which is similar to original plaster in feeling. The material may tend to shrink when applied in this fashion; dry it relatively slowly and without extreme heat. Minor cracks can be repaired once the durabond has dried. The surface may then be painted.

Although decorative plaster elements like medallions or cove moldings should probably be repaired

or installed by a professional, there are a few alternatives within the reach of the owner-preservationist. Simple plaster covework can be nicely duplicated by using combinations of wooden moldings. After they have been attached to walls or ceilings, apply two or three coats of gesso in acrylic form, which can be purchased at art supply stores. The gesso will re-create the feeling of plaster. Reproduction plaster decorations are being prefabricated in a number of small shops throughout North America. It should be relatively easy to acquire stock items or duplicates of the originals in your house. The costs are surprisingly low.

It is now time to consider the decoration of the interior. It is important to recognize from the outset that latex paints, while they are very popular and easy to work with, will not be compatible with the lime in old plaster; chemical reactions will cause bubbling or pitting. In the past, paints were always applied with a brush. Rollers leave a slightly raised, "orange peel" effect that is inconsistent with the texture of plasterwork.

Paints should adhere to historical compositions wherever possible. Early paints in North America

ABOVE

This nineteenth-century wall sconce highlights the irregularity of the early plasterwork.

OPPOSITE, ABOVE

The first stage of creating a cove is to lay on a base coat of plaster.

OPPOSITE, BELOW

The finish coat is then laid on by applying plaster to an appropriately contoured template, which is then pushed along the form. Fresh plaster is left behind, providing the appropriate cove detail.

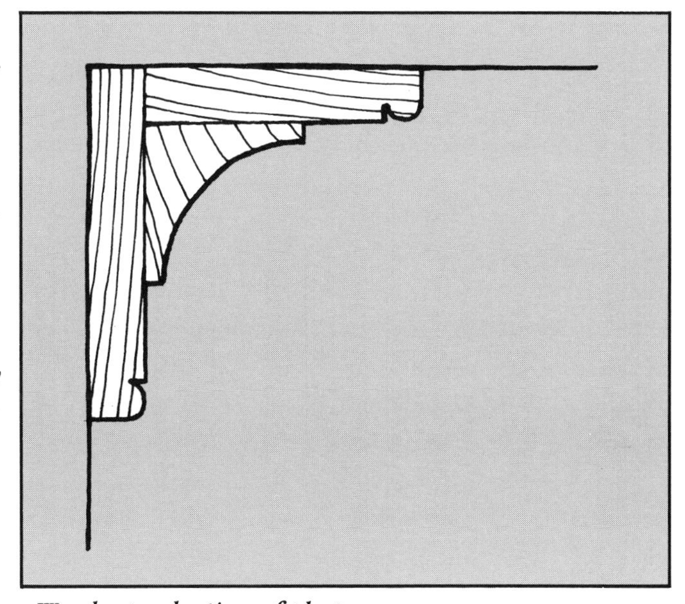

Wood reproduction of plaster

would be made up of pigment, binder and drying agent. For example, pigment would be mixed with linseed oil and japan dryer. Lead-based paints were common. Today, commercially available alkyd paint is the only ready-made product recommended for use on original plasters.

In the past, distemper was commonly used as a wash over plaster. This material is extremely difficult to buy commercially these days, but you can make your own by following the recipe given in Chapter 9, where you will also find a list of other wall coatings.[1]

The plaster surface must be thoroughly prepared before any painting is done. Over time, original surfaces may have been plainly painted, decorated or wallpapered. In many cases all three have occurred. Damaged areas will have been repaired previously, including the filling of nail holes, etc. If wallpaper is to be removed, use a professional-type steamer that can be rented or purchased. Liquid "wallpaper-removers" available at paint outlets do seem to work, but they are much slower and much messier. Carefully wash the walls and ceilings with lukewarm water. Coat all rust or heavy staining with two coats of shellac to act as a sealant. (Make sure you deal with the cause of the stains.) The surfaces are now ready to be painted or re-papered.

The use of wallpapers is often misunderstood. As in any preservation project, modern tastes should not compromise historical integrity. The application of paint in period interiors where papers were originally used falsifies the builder's initial interpretation of the mood, effective light and general proportions of the room. The appearance of wallpapers in Europe can be traced back to the beginning of the silk trade with China in the eleventh century. Coverings made of rag or leather in imitation of the fantastically expensive silk quickly became popular. Individual sheets were printed and pasted on the wall, and the overall effect would be much the same as we know today.

Wood paneling as a wall covering was common in early North American domestic architecture, but the limited use of wallpapers has been documented from the early 1700s. However, it was not until the 1800s that wallpaper became available for the common man. Contrary to some contemporary opinions, vast expanses of off-white wall would not be viewed with favor. In many cases, highly patterned papers were preferred because of their elegance and as proof of affluence.

If you choose to duplicate the original feeling of a wallpapered room, research is the best answer. Attempt to steam off large pieces of wallpaper around covework or from the general wall area. Separate the layers of paper you have removed by steaming (an ordinary kitchen kettle will do) from the underside of each layer. Spread the layers out on plastic screen or blotters. By researching wallpaper use in your area, you will be able to ascertain the approximate date of each application. Period wallpapers may be obtained from many North American and European sources.

On the other hand, if the outer layer of paper is in good condition and of the appropriate period, a simple dusting with a vacuum cleaner and a soft bristle brush should rejuvenate it. More serious dirt can be lifted with a draftsman's eraser, the kind that comes in a bag.

The misuse of wallpaper styles for given historical periods is a classic problem in North American preservation. What today's taste views as of the period

OPPOSITE, ABOVE

The fireplace surround in this Scottish or Irish one-and-a-half story house uses simple joinery.

OPPOSITE, BELOW

Paneling was an early form of interior finish. This "frontier" example is at Louisbourg, Cape Breton Island. First half eighteenth century.

and attractive may be completely out of context with the original. In fact, the original may be totally offensive.

The answer to this and similar problems encountered in the interior is compatibility linked with historical accuracy. Very few of us would be purist enough to live willingly with some of the florid wallpapers of the late 1890s, but less violent patterns from the same period are available. It really becomes a matter of understanding in the mind of the preservationist. The interior decorations of the past had an elegant harmony of their own that is frequently difficult to describe and time-consuming to duplicate completely. But the result is definitely worth the effort.

It is easy enough to say that period paint schemes might involve butting together primaries like red and green or ochre and blue in the same room, but it is difficult to convince the novice that such combinations can work beautifully unless the results are actually seen. The paints made for today's pastel interiors are pale and washed out compared to the deep, rich colors of the past. Such colors can still be found, but their use requires research and some bravery at the outset.

The key is harmony, part of which is achieved through concentrating on historically correct details. Interior hardware is a small but significant aspect of the overall feeling. Glass doorknobs from the 1890s would be totally out of place in an 1840 stone farmhouse. Equally, wrought-iron thumb-latches in the dining room of an 1890 brick town house would be discordant. The best guides for interior decoration, once the plaster has been preserved or re-created, are thought, patience and research.

ABOVE

Careful removal of later wallpapers (shown on the left) has revealed an earlier paper consistent with the overall restoration of this 1825 house.

OPPOSITE

Stenciling, throughout the late eighteenth and early nineteenth centuries, added a charming decorative element to both interior finishes and furnishings that was previously unknown to the settler of modest means.

ABOVE, LEFT

A series of hooks, bolt assembly and masonry hangers for use in restoring early nineteenth-century interiors.

ABOVE, RIGHT

A nineteenth-century box lock.

OPPOSITE

A delightful restoration, c. 1890.

CHAPTER SEVEN

Masonry in the Landscape

Semi-elliptical cobblestone fans sweep through this courtyard.

THROUGHOUT THE HISTORY of North American domestic architecture, the use of stone and brick in the surrounding landscape has enhanced and accented the house. The restoration of the garden or yard is as important as the rest of the masonry building, because the treatment of the landscape can create or destroy the mood of the past at a single glance.

The earliest landscaping forms in North America were strictly geometrical. Following the long European tradition, a series of built-up rectangular or square planting beds would be laid out with paths surrounding them. Pathways were often made of brick placed in a decorative pattern, herringbone being a common example. At this point, coastal settlements were the norm, and both sand and crushed shells were used as path materials. Shells were also burnt to make lime for mortar, and it seems likely that shell paths would become fairly solid over time as feet compacted them and the rain poured down. Absolute balance in proportions was the ideal in these early gardens, and the entire area would be fenced with wooden pickets or stakes.

The dooryard was a later development. It was basically an outdoor workshop enclosed by fences and out-buildings. Apart from stone easily collected from nearby fields, most fencing was made of more workable and available wood. Within the dooryard an area or areas would be further fenced off where vegetables, herbs and fruit would be grown.

Such rural gardens were carved from the rough landscape, but the towns and cities were faring somewhat better by the mid-1700s. At that time both New York and Boston boasted cobblestone streets, although most smaller settlements still had dirt roads. Brick or stone walls in the European manner were occasionally seen, but stone more so than brick. In populated settlements masonry walls were sometimes built as firebreaks, although their usefulness around the garden was open to question. John Lawrence observed in his *A New System of Agriculture* (1776):

> Walls are some defence where they are tall and the garden Little. But otherwise they occasion Reverberations, Whirles, and currents of wind, so they often do more harm than good. I should therefore choose to have the Flower Garden encompassed with hedge.[1]

Along the eastern seaboard at least, the harshness of pioneer life was lessening by the mid-1700s. Houses now contained rooms of leisure, and the complementary "parlor garden" came into existence. Flowers and shrubs were grown in this area, which was often no more than the width of the house and two-thirds that deep.

In Europe, a more natural or "wild" form of landscaping was coming into vogue on large estates, but the earlier geometrical type was still common until the late 1800s. Throughout the nineteenth century, J.C. Louden and A.J. Downing were as influential in garden layout as they were in architecture. To some degree their ideas of the less formal garden were founded on the earlier changes in European attitudes.

Louden was instrumental in bringing landscaping to the attention of the masses in England through his designs for public parks. As he himself admitted, North America was the land of Everyman, and its generally rural nature did not as yet require the public gardens he built for the escape of the metropolitan, industrialized English worker.

It was left to A.J. Downing to found the American public park. As in architecture, his methods were frequently copied. His fascination for things Gothic suited the mood of the Victorian era. In designing domestic forms, he favored well-situated houses with lawns surrounding them. Flower beds were circular or arched. Stone and brick were often used for pathways and to build temple-like garden houses that bordered on the Rococco in style. Cast-iron fences frequently surmounted bases of cut stone in the North or brick in the South.

The preservation of landscaping features must be done within the context of the period in which the house was built. As in all period elements, inade-

ABOVE

Brick walling with limestone caps and decorative ornamentation. Note the use of both diamond and random flagstone pavers and the uncut stone border edging.

RIGHT

A polychromatic; herringbone pathway is a suitable adjunct to a stone or brick house.

quate maintenance or simple old age will lead to problems in the masonry components.

Brick or stone paths will become irregular or the units will be loosened as drainage areas become overgrown. In turn, the base layer is washed out. Failed masonry joints will magnify the problem as even more moisture enters the space. Paths in close proximity to buildings may hide foundation problems in the structure itself. The buildup of vegetation, although picturesque, makes walkways hazardous when wet, and the invasion of root systems will eventually lead to the breakdown of masonry units. As in the house itself, moisture penetration will result from the ravages of the freeze-thaw cycle; porous bricks will spall and crumble, stone will crack.

The wear of time is a major reason for acquiring an old house in the first place, but wear and damage are two different things. A century or so of visitors to your house will leave their indelible and wonderful mark on a stone step. Unless the wear is causing problems, best leave it, and enjoy it. But if structural integrity has been affected or moisture has penetrated between the step and surrounding elements, remedial action must be taken.

Contemporary forms of path maintenance may themselves be the most damaging factor. Snow clearance by machine can chip masonry edges and mar surfaces; brickwork may be crushed altogether. Salt as a de-icer will cause breakdown in the pathway and in nearby house walls.

Changes in life patterns may require new pathways, while major rebuilding of old ones may be necessary. Follow the illustrations for acceptable methods. When building with brick, new units can be purchased that are specifically designed for the purpose. The grade will probably be specified as "SW." Be sure to examine the different colors available and choose one that is most compatible with the original. Build with traditional jointing methods in mind. If there are no original models on the property, examine vernacular styles in the immediate neighborhood. If in doubt, keep it simple.

Two styles of laying masonry paving are mortared and mortarless. The first requires a rigid base or slab, and in areas where deep frost penetration occurs, it can prove costly. Unless adequate attention is paid to the base, frost heave will ruin the work in the first year. The bricks or stones are laid in a bed of mortar, and the joints are pointed with the same material. In general, this method should be avoided.

The second style will allow you to lay paving accurately, effectively and cheaply. Frost heave is not a problem. Stake out the area and set up lines where the path will run. If sod is to be removed, cut along the lines and take it up slowly. Transplant it in a needy area. Scoop the topsoil away to a depth of about three-and-a-half to six inches. To prevent eventual breakdown, line the edges with bricks standing on end. The most important element in mortarless paving is the subsurface. Install it as in the illustration; make sure it is level and adequately compacted. When building with brick put down a layer of roofing paper or polyethylene; omit this step with stone. Lay the brick or stone in the desired pattern, leaving approximately one-quarter of an inch between bricks and following the natural irregularities of the material when using stone. Mix up a batch of dry mortar with one part Portland cement, two parts lime and eight parts sand. Spread the unmoistened mixture over the path and brush it into all the joints and voids with a broom. Remove any dry mortar from the faces of the units. Lightly sprinkle the area with a fine mist of water. This will cause the mortar to set up into a very weak bond flexible enough to allow constant wear.

The mortarless method can be effectively implemented when resetting original pathways. Before taking the old path up, however, sketch or photograph it so reassembly will be historically accurate.

Gravel pathways, on the other hand, can be easily laid by following an 1831 plan given in *Five Thousand Recipes in the Useful and Domestic Arts*:

The bottom should be laid with lime rubbish, large flint stones, or other hard matter, for eight to ten inches thick, to keep the weeds from growing through, and over this the gravel is to be laid six or eight inches thick. This should be laid rounding up in the middle, by which means the larger stones will run off to the sides and may be raked away; for the gravel should never be screened before it is laid on. It is a common mistake to lay these walks too round, which not only makes them uneasy to walk upon, but takes off from their apparent breadth. One inch in five feet is sufficient proportion for the rise in the middle, so that a walk of twenty feet wide should be four inches higher at the middle than at the edges, and so in proportion. As soon as the gravel is laid it should be raked, and the large stones thrown back again; then the whole should be rolled both lengthwise & crosswise; and the person who draws the roller should wear shoes with flat heels, that he may make no holes, because holes made in a new walk are not easily remedied. The walks should always be rolled three or four times after very hard showers from which they will bind more firmly than otherwise they could ever be made to do.[2]

Stone and brick walls used in landscaping suffer the same problems as any masonry structure, although they do not endure the same thermal stresses experienced by the masonry building heated from within. Rebuilding and pointing techniques are the same as those outlined in Chapter 3.

One unique problem occurs when planting boxes have been set into a wall. Poorly maintained boxes allow roots and moisture to penetrate the core and joints of the wall, breaking the masonry apart. The effect can be charming, especially when the boxes are filled with the older flower varieties, and once the damage has been repaired there is no reason why they cannot be used.

High, narrow walls should be capped with a lime-based mortar that contains a fairly high proportion of white Portland. This type of mortar is used on chimneys and parapet walls. The recipe will be found in Chapter 9. It is extremely important that the mor-

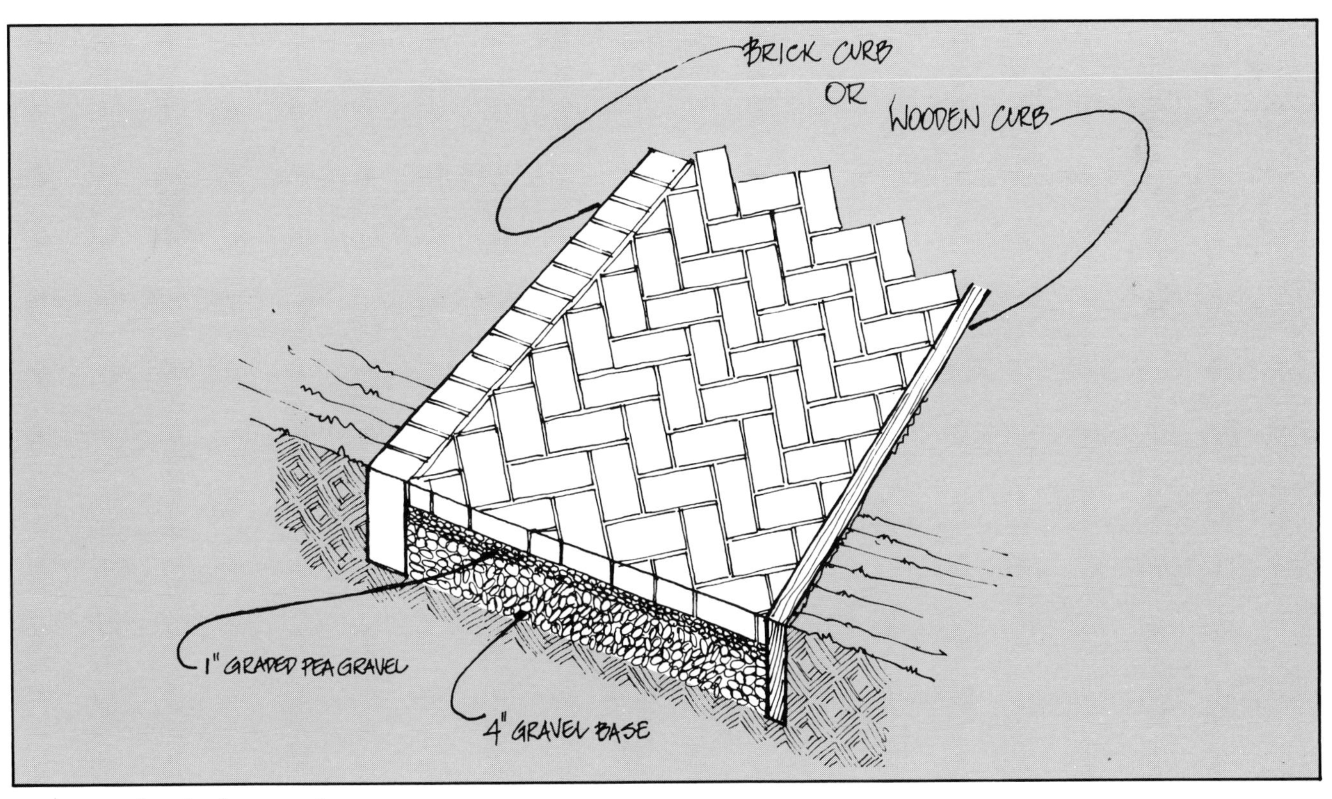

Laying a herringbone path

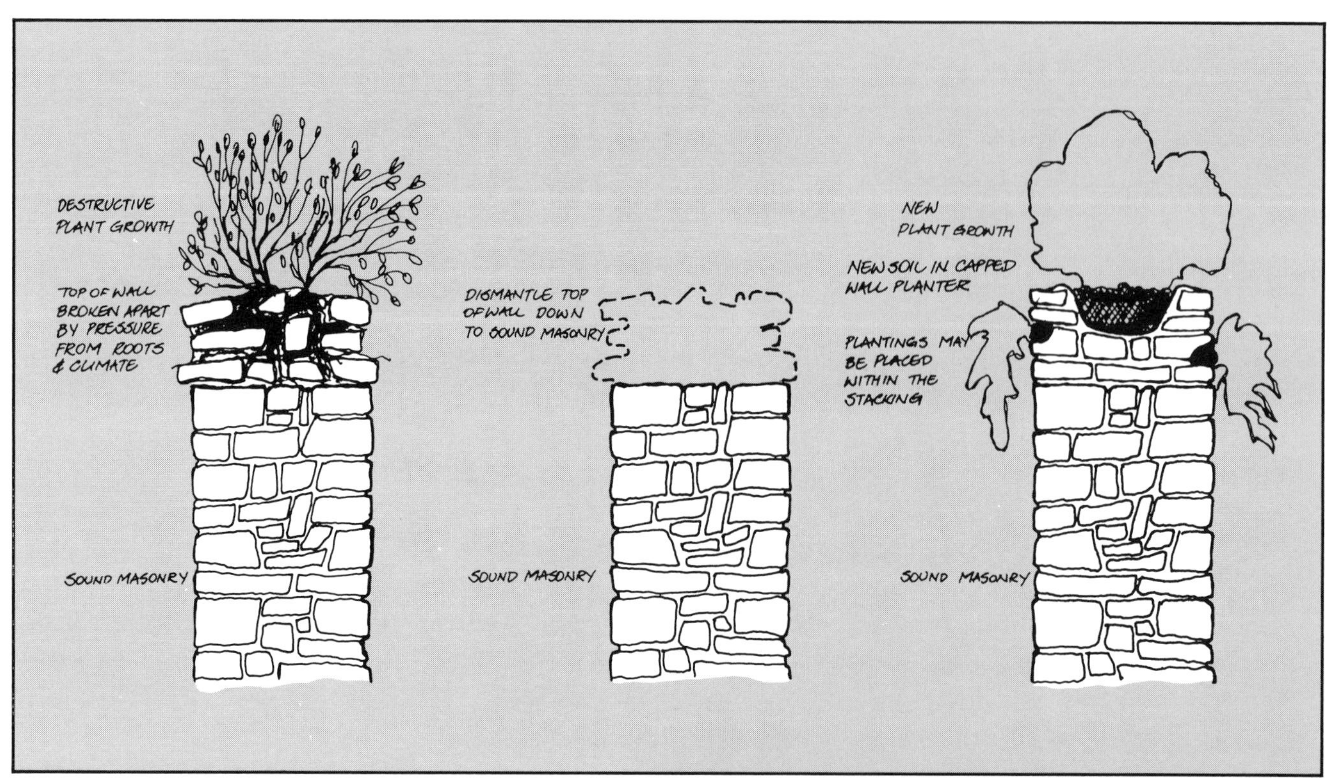

Repairing landscape walls

Capping an exterior masonry wall

tar be applied to a moist wall and then vigorously troweled into a dome shape to allow adequate water runoff. Moisture in the mortar will cause spalling as it sets unless properly troweled. A very effective water-impervious membrane can be applied to the cap after the mortar has set. Called Neolon, it was originally developed for boat decking. It will absorb normal masonry movement without splitting and maintain its elasticity at extreme temperatures.

The more ambitious preservationist may wish to build entirely new rubblestone or brick fencing. Footings and building techniques are the same as in Chapter 4. Capping treatments will depend on the vernacular style used during the period in your area. A few photographs taken in the neighborhood should help you make the right decision.

About the best way to sum up this chapter is by relating a recent experience. I had stopped in to visit some friends who were busily restoring their 1820 stone house. It was fall, and they had spent part of the summer working on the landscaping. Their home was in a village with many historic buildings, but most had been "renovated" in one form or another. They were concerned because an acquaintance was coming to call for the first time. She was late, and they thought she had missed her way. When she did arrive, they apologized for their poor directions, but she replied, "I had no trouble finding the place. I spotted it from four blocks away. It's a little corner of the nineteenth century."

1

2

3

4

5

6

7

8

9

10

11

12

13
14
15
16

17

18

19

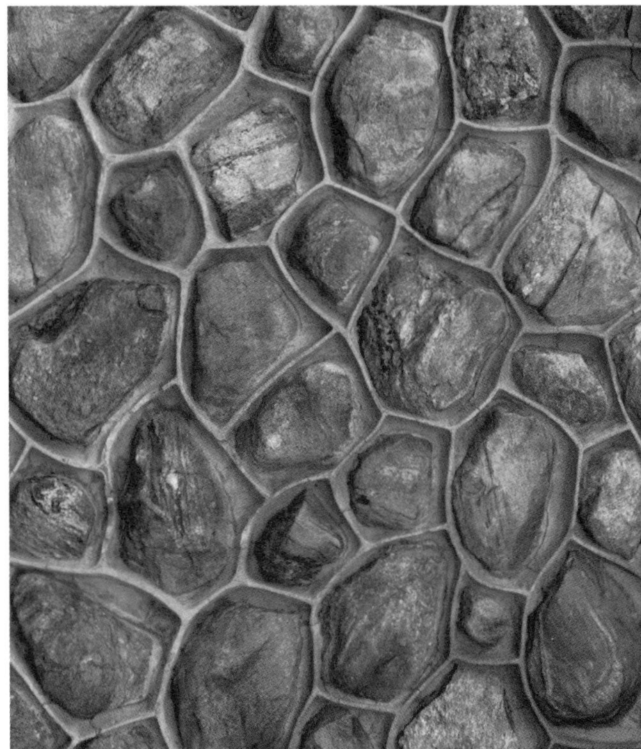

20

21

22

23

24

25 *26*

27 *28*

1. *A classical Georgian entranceway. Note the rectangular flagstone walkway and ashlar steps.*
2. *An elegant blend of period and contemporary elements, c. 1900. The white markings on the steps are salt deposits from winter snow removal.*
3. *This colonial revival entranceway is particularly effective, with its cut stone walkway and stairs framed by Neoclassical urns, and its dramatic use of shrubs and vines.*
4. *An inviting entrance to a turn-of-the-century home with a cut stone walk and limestone stairs.*
5. *A Victorian treat: the porch has been glassed-in, making a usable space for more than one season. Note the marriage of plantings to the facade.*
6. *This Georgian house is framed by brick pillars with limestone caps and bases; limestone edging delineates the grass and drive.*
7. *A simple marriage of landscape and entranceway in this Victorian townhouse.*
8. *Ashlar walling and pillars, c. 1930. The decorative stone capping and ornamentation are accentuated by the rampant ivy foliage.*
9. *Cut limestone walkway, pillars and base course with ornamental iron gate and fencing: strong, elegant and permanent.*
10. *Gateway with brick walling and pillars with limestone caps and base.*
11. *Stucco wall and pillars with a cut limestone cap, c. 1930.*
12. *A dramatic Italianate cut limestone gatepost.*
13. *A gateway with a terra cotta brick path, framed with stucco walls and a cut limestone cap.*
14. *Brick walling with lateral brick capping; cut flagstone pavers, steps and edging. The proportion of the wooden gates and pillars draw one to the entranceway while the hedging softens the overall mass of the wall.*

15 *A cobblestone arched entranceway.*

16. *A cobblestone garden shed built into a cobblestone perimeter wall, c. 1930.*
17. *This brick walling with a limestone cap and ornamental iron fencework is especially effective with the dramatic detailing of the adjoining stucco.*
18. *Brick walling and pillars with a cut limestone cap, base course and finial.*
19. *Random cobblestone walling with raised pointing, which is used as a decorative feature as well as for weatherproofing.*
20. *Detail of random cobblestone walling.*
21. *A coursed dry stack rubblestone wall capped with mortar and horizontal stonework.*
22. *A dry stack fieldstone wall, c. 1850.*
23. *A dry stack random rubblestone wall and gatepost.*
24. *An ashlar fence post.*
25. *A semi-circular decorative arch from a commercial building in an urban area that is now used as a decorative feature in a garden landscape.*
26. *A column from a demolished commercial structure now soars skyward in a garden setting.*
27. *A splendid pair of Neogothic gateposts. Especially notable here are the brick walling returns and gravel walkway.*
28. *A detail of the top of the brickposts with terra cotta decorative insets, caps and finials.*

CHAPTER EIGHT

Recycling Non-Domestic Buildings

Classic contemporary design is used in this converted warehouse.

THE WIDESPREAD ENTHUSIASM for the architecture of the past has branched out from historical domestic forms to include the innovative recycling of non-domestic structures as living space. Warehouses, barns, post offices, railroad stations, schools and mills have all been successfully adapted to new uses. The analysis and maintenance of these buildings is the same as in the period masonry house, but unique creative opportunities are frequently presented in the conversion of interior space. A very real benefit of dealing with such industrial or semi-industrial structures is that twentieth-century mechanical systems and insulation can be introduced without harming period details.

The development of the house itself stemmed from the need for shelter. A roof and walls were constructed and a space enclosed. As the daily battle for existence lessened in intensity, the concept of more than one room evolved. A sleeping niche or loft was the first division of the basic one-room structure. One seventeenth-century North American writer remembered,

> I then betook me to my apartment, which was a little Room parted from the kitchen by a single bord partition.[1]

By the eighteenth century basic partitioning of the one room had become relatively common. As the century progressed and homes evolved to mirror their owners' establishment in the community, more rooms or even whole houses were built using the originals as ells or wings. The parlor or leisure room became part of the basic design. This was not an immediate or complete process. Even as late as 1900 rural areas were still in transition from basic shelter to more comfortable living space.

As these events were taking place, buildings needed to contain industries and services were also being constructed. Often the same materials and methods used by the housebuilder were employed, albeit on a much larger scale. Rubblestone, for example, was commonly used for barns and mills in North America. These two types of buildings in particular are frequently discussed with a degree of historical and romantic affection. Their survival is a primary statement of the rural foundations of early settlements.

Barn and house were often built as one, and perhaps it is this early association that leads us to consider the space within its walls as a potential living capsule for today. Mills seem to evoke the same response. Rather than a historic domestic interior in need of restoration to return it to its former glory, the open space offered by barn or mill gives the owner carte blanche in design.

Interior elements, such as support systems, should be used as part of the overall breakup of form. A barn with a kingpost truss system can be elegantly redesigned keeping this element in mind. Interior harmony is a significant aspect in any recycling, and in this case rigid foam insulation may be employed on the roof and cladding applied over it.

When outlining the social and domestic needs of your family, the best approach is to list the physical requirements — kitchen, studio, office, bedrooms, bathrooms — and the amount of space required for each one. If the area within the structure is quite large, only a portion of it may be needed as a living capsule. One couple of my acquaintance owns a very extensive rubblestone mill in a lovely rural setting. Their needs are met quite handsomely by using one part of one floor. There are several others available, one of which is used for roller skating.

Once interior requirements are understood, consider the visual elements offered by the landscape. Note how windows admit light and a view of the land. In northern climates, it would seem to be

OPPOSITE, ABOVE
Existing but unused factory space is ideal for urban renewal.

OPPOSITE BELOW
Contemporary forms accentuate this urban loft renewal.

This home/studio is a fine example of space reuse. This building is part of a former warehouse complex in a downtown area.

common sense to use the southern portion of a large building.

Mills and barns are early non-domestic forms. The Industrial Revolution added substantially to this initial stock. In the urban areas of the nineteenth century, brick and stone industrial buildings, often in combination with cast-iron posts and joists, began to create whole neighborhoods themselves. I vividly remember one large complex of brick buildings originally designed as a woolen mill. Sadly it has been torn down, but at one point in the nineteenth century it covered almost five acres. It was a beautiful statement of the physical shape of early industry that should have been saved for its own intrinsic merit, but the point here is that it could have been recycled and effectively used. The cost per square foot for recycling such buildings is significantly lower than for new construction.

What is now downtown in many centers is frequently made up of early warehouse forms. The philosophical and design approach to these buildings is no different than with a barn or mill. Proximity to the city core and reasonable purchase or lease prices make them attractive to the contemporary preservationist. In a barn or mill, truss systems might be maintained as decorative elements integral to the overall interior layout. In Victorian industrial buildings, period mechanical systems lend themselves very well to exposure as textural and sculptural elements.

Modern mechanical systems, the requirements of structural support and insulation, on the other hand, must all be carefully analyzed if successful conversion of space is to be achieved. The exciting thing about recycling these buildings is that there is vast scope for the imagination of the designer. Interiors can be decorated along traditional lines using plaster and period trim, or they might take a completely modern turn.

CHAPTER NINE

Tools, Mixes and Methods

Owning a stone house has always been something of a status symbol, and imitations were built for nearly two centuries. This early wooden facsimile of ashlar is suffering from paint deterioration. The surface is being prepared for new paint.

ANYONE WHO ATTEMPTS to preserve a brick or stone house will need to understand many trades and disciplines. The original builder had probably spent many years as an apprentice before attempting your house. Or, in the case of a rudimentary building, it was the result of the combined help and talents of neighbors and friends. The masonry arts are a complex series of disciplines that even the most dedicated amateur may never master, but the various preservation techniques and building processes in this book can be performed at an acceptable level by the novice. There is such a thing as being manually adept, and it is a gift. Your first ten minutes with mortar and trowel will be disheartening, but the mason's art, like any other profession, takes practice and patience.

Successful preservation is a function of the ability and knowledge of the preservationist. Without the right tool, the job cannot be done properly. Without the correct mortar or coating, the house will suffer aesthetically, historically and structurally. Before a task is performed, it is fundamental to be able to estimate the amount of material needed; too little is frustrating, too much is wasteful. Through illustration and explanation, this chapter will help you grasp the fundamentals of the preservation process. Some of the materials listed here can be difficult but not impossible to find if you are persistent. Wherever possible, sources or alternatives are given, although it is impossible to cover all areas.

Tools

It is worth repeating that the quality of your tools will affect the quality of your work. Buying fine tools may seem like a needless expense at the outset, but my experience has shown quite conclusively that cheap tools are false economy.

Stone and Brick

- Chipping hammer
- Claw hammer
- Hammer (two-pound)
- Blade chisel
- Cold chisel
- Mason's trowels (six-inch and ten-inch)
- Pointing trowels (one-eighth-inch and one-quarter-inch)
- Mason's hawk
- Plumb bob
- Mason's string (it doesn't stretch)
- Metal shears for brick ties
- Chalk
- Long mason's level
- Short line level that attaches to mason's string
- Tape measure
- Graduated rubber pails
- Shovel
- Mixing trough or mortar mixer
- Wheelbarrow
- Mason's hoe
- Hose with spray nozzle
- Whitewash brush
- Unprinted, undyed burlap

Plaster

- Laying-on trowel
- Hawk
- Homemade smoothing tools: for small areas, a piece of 12″ × 6″ × $\frac{5}{8}$″ smooth pine with a firmly attached wooden handle; for large areas, a 36″ × 4″ × $\frac{5}{8}$″ piece of pine with two handles.

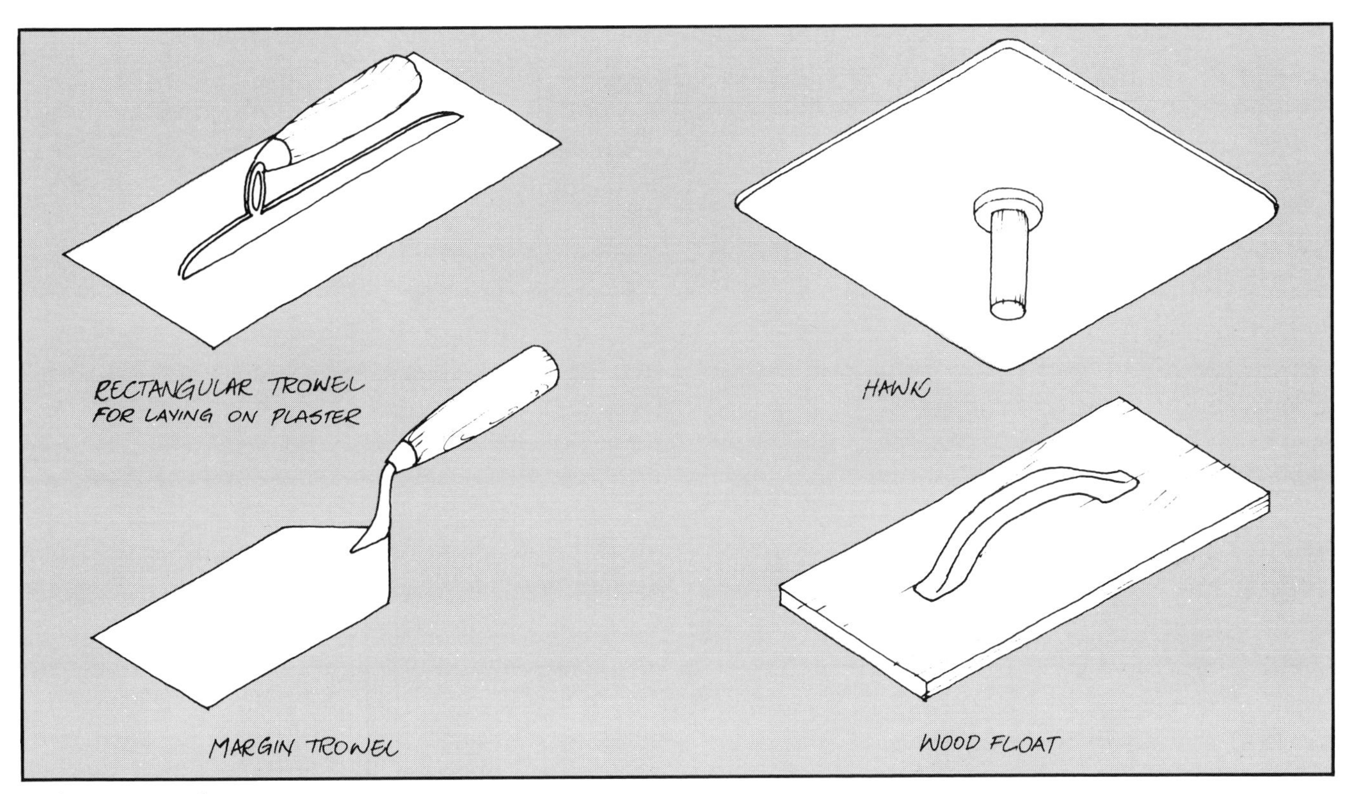

Plastering tools

- Small (three-inch) mason's trowel or wide putty knife for corners
- Shovel
- Mixing trough or plaster mixer
- Buckets
- Homemade work board: a 4′ × 4′ × $\frac{3}{4}$″ piece of plywood supported on sawhorses to hold the mixed plaster
- Stepladder
- One section of scaffolding with cups under the legs to protect the floor
- Thick plastic sheeting or newspaper to cover the floor and prevent scraping later

Plaster Repair

- Electric drill with a variety of bits
- Minimum two-inch-long screws with narrow shank and coarse thread (plated)
- Washers that screws will fit
- Variety of scrapers
- Can opener (punch type) for scoring cracks
- Buckets
- Spackle or crack filler
- Medium and fine sandpaper
- Plastic ketchup or mustard bottle for squeezing filler into deep cracks
- Whitewash brush for wetting surfaces
- Sponge or cloth for smoothing filler
- Masonite mortar board (approximately ten-by-ten inch) with thumb-hole or handle
- Mixing trough (a baby bath is cheap, sturdy and easy to clean)

Wallpaper Removal

- Mat knife for scoring
- Large sponges for wetting

- Garden weed sprayer (good for wetting selected areas)
- Chemical wallpaper remover (optional, mix with water)
- Wallpaper steamer (can be rented, very fast)
- Buckets
- Stepladder
- One section of scaffolding with cups under the legs to protect the floor
- Newspaper to cover the floor and prevent scraping later

Basic Safety Equipment

- Hard hat
- Work boots (steel toe models are a good investment)
- Work gloves
- Goggles or glasses

Rented Power Tools

Traditional building methods depended on hand-held and hand-propelled tools. One of the better inventions of the twentieth century is the local rental center specializing in power tools that are far too expensive to buy but make tedious jobs at least bearable. A number of these can be employed by the novice in masonry and plaster work.

A mortar mixer resembles a plaster mixer in form and does basically the same thing, except that the mixing paddles are more drill-like and larger. Before you rent or use either kind, make sure the mixing box has been cleaned after the last job. Otherwise, operation will be extremely difficult. A buildup of mortar or plaster around the drive shaft will slow the machine down considerably. Every time the mixer is used, lubricate the nipples on the shaft with a grease gun. A rental agent who likes his equipment will gladly provide you with one. (Sand and lime are not compatible with machinery.) Both types are mounted on wheels for easy positioning at the site. Allow adequate room for a wheelbarrow to come up alongside and turn around. Spills always occur, so put plywood or planks under the dumping area.

A cement mixer, on the other hand, has a conical drum mounted on a stand. The cone is divided by small mini-walls that hold the aggregate in place sufficiently to separate and mix it when the machine is turning. Mortar and plaster mixers dump from the side, but the concrete variety dumps from the end. The main reason to use a cement mixer when any amount of crushed stone is present has to do with the fact that the aggregate will jam the paddles and seize the shaft of the other varieties.

Mortar, plaster or concrete ingredients should be mixed dry first, then water should be added to the desired consistency. Mortar and plaster mixers are gas-driven; the concrete mixer is powered by gasoline or electricity. Be prepared with outlets, power cords and supplies of fuel.

Cement mixer

Mixing these materials by hand is almost as tedious as chipping out masonry joints. The best chipping tool I have found is the Ingersoll Rand scaler. It is light, very strong and long lasting, and the handle is shaped for human beings. Quite literally, all the cheaper models will fall apart, even on small jobs. The power source for a scaler is a compressor. Most rental shops will carry the Ingersoll Rand brand, as well as the compressor and its hoses. Check the size of the masonry joints and pick a bit of suitable width.

A power trowel is a gas-powered item that looks very much like a garden roto-tiller, but instead of rotors it has paddles that swirl around parallel to the ground. When working on concrete floors, this device can be used to advantage during the finishing stages. The concrete should be stiff but moist enough to work. The blades come off the trowel, so take the machine into the area first and then attach them. In effect, the machine becomes your arm moving at about five miles an hour. The paddles should not dig into the surface but skim over it. Practice will be needed. Unless you live beside the rental store, rent two machines, because one almost always breaks down halfway through the job.

A power compactor resembles a pneumatic drill with a pad at the end. Run by a compressor or electricity, it is used to pack down bases for masonry or landscaping. Nothing does the job quite as well.

Heat Guns

These are electric guns that heat up to remove paint. Some have blowers within the unit and look something like large hair dryers. Others are merely pans heated by coils. I have used both kinds, and both are recommended, although the blower types offer slightly more control.

Heat guns are particularly useful when original finishes are being restored, because when used carefully they will not disturb period paints. Apply the heat from a greater distance than in complete stripping, so it does not darken shellacs or varnishes.

Scaffolding

Scaffolding is easily rented for a nominal fee and is recommended for all but the most stepladder-type jobs. A standard section is made up of two ends five feet wide, fifteen feet high and seven or ten feet long, depending on the type of scissor brace that holds them together. Make sure the clips for holding the braces are part of the end assembly; loose ones will get lost or fall on you while the sections are being put together.

The illustration on page 68 shows how to position the built-in ladders when erecting scaffolding. The planks laid across the metal frame should be two-by-ten inch spruce twelve feet long. A five-board covering at the top of each section is a general rule of thumb. Safety rails should be employed.

In urban areas, close proximity to sidewalks often makes "sidewalk scaffolding" desirable. The general dimensions for this type are six feet wide, nine feet high and seven or ten feet long, depending on the brace. Sidewalk scaffolding must be covered, not only with planks, but also with some material (chipboard) to prevent debris from falling through onto pedestrians. If you plan to build higher than one frame, adaptors will have to be rented to accept the five-foot-wide standard section that should be attached above the wider first section.

Scaffolding must be even and secure. If it is uneven, sections will not fit together after the second frame in height. When ordering, make sure you get adjustable legs if you are working on uneven terrain. The weight of stone or brick on an uneven or insecure scaffold may cause it to topple, resulting in serious injuries. Any doubts about security can be resolved by renting cross-braces that go from the scaffold to the wall on which you are working. In general, sloppy scaffold building is dangerous to your health.

If more than four frames and braces are needed, pay to have them delivered. Scaffolding is heavy, cumbersome and not prone to fit on economy car roof racks.

Mixes

The materials listed for the mixes given here should be of the best possible quality. Mortar should always be made using clean tools and drinkable water. Sand should be of an even grade and free from impurities. Protect it from rain, because sand absorbs a lot of moisture when it sits out and dampness will result in sloppy mixes. Only non-staining white Portland cement (sometimes called white medusa) is the correct type unless otherwise specified. Use hydrated lime for masonry purposes. All mortar for brick and stone construction should be moist enough to stick to a trowel but dry enough not to cause undue smearing or staining.

Purists may want to hydrate their own lime, but it will not be as effective as the commercial variety. Traditionally, lime was slaked by mixing up a solution of sixty percent calcium oxide and forty percent water and letting it soak for a few days.

Traditional Mortar

- Three parts sand
- One part slaked lime
- Water

Good Historic Duplicate

- One part white Portland cement
- Three parts hydrated lime
- Twelve parts clean sand
- Water

Mortar for General Use and Below Grade

- One part white Portland
- One-quarter part hydrated lime
- Three parts sand
- Water

Mortar for Chimneys, Parapet Walls, Capping

- Two parts white Portland
- Three parts hydrated lime
- Six parts sand
- Water

Mortar for Solid Masonry, Load-bearing Walls

- One part white Portland
- Two parts hydrated lime
- Nine parts sand
- Water

Mortar for Retaining Walls

- One part white Portland
- One-half part hydrated lime
- Four-and-one-half parts sand

Mortar for Solid Masonry Foundation Walls

- One part white Portland
- One part hydrated lime
- Six parts sand

Joint Colorants

- White — white Portland cement
- Dark blue, blue-gray — Germantown lampblack, black oxide of manganese, or carbon black
- Brown — burnt umber, or brown iron oxide
- Brown-red, brick red, pink, salmon — red iron oxide or powdered brick dust
- Red sandstone, purple-red — Indian red
- Buff, cream, yellow — yellow ochre or yellow iron oxide
- Blue — ultramarine blue

Tuckpointing

- One part crushed red brick
- Four parts lime-based mortar

Concrete for Footings

- One part ordinary Portland cement
- Three parts sand
- Four parts aggregate (gravel)
- Water

Concrete for Basement Floors

- One part ordinary Portland cement
- Two-and-one-half parts sand
- Three-and-one-half parts crushed stone
- Water

Cold Cement, Richard Neve (1703)

"Take half a pound of old cheddar cheese, peel, grate very small, put into pot. Then a pint of cow's milk, let stand all night. Then get whites of 12-14 eggs, $\frac{1}{2}$ lb. of best unslaked or quick lime, and beat to powder in a mortar, then sift it through a fine-hair sieve into mixture, stir well, breaking knots of cheese, then add whites of egg and temper well together. This cement will be white — add brick dust if brick color required."[1]

Exterior Stucco

- First two coats: -one part white Portland
  -one-tenth part hydrated lime
  -three parts sand
  -water
- Finish coat: -one part white Portland
  -one-tenth part hydrated lime
  -two parts sand
  -water

A.J. Downing's Stucco

- One part lime
- Two parts sand
- Water

Use two coats; apply the second before the first is dry.[2]

Historic Stucco Coating

- Primer — a little red lead mixed with linseed oil
- Second coat — mix of white lead, a small portion of red lead, linseed oil and a little spirit of turpentine
- Third coat — white lead mixed with linseed and turpentine in equal parts
- Fourth coat — half white lead and half turpentine and linseed
- Fifth coat — turpentine and flatting

Stucco Mastic, Patented 1815

Linseed oil boiled with litharge and mixed with porcelain clay finely powdered and colored with ground brick or pottery. Use turpentine as a thinner.[3]

Dutch Tarrass for Foundation Parging

- First coat: -two parts lime
  -one part plaster of paris
  -water
- Finish coat: -one part lime
  -two parts well-sifted coal ashes
  -water

Adobe Whitewash

- One part ground gypsum rock
- Two parts clay
- Water

A.J. Downing's Whitewash

"Slake one-half a bushel of fresh lime in a barrel, fill the barrel two-thirds full with water and add one bushel of hydraulic cement or water lime. Dissolve in water and add three pounds of sulphate zinc. This should be the thickness of paint; it can be improved by stirring in a peck of white sand. The color is pale stone, nearly white."[4]

An 1840 Whitewash Recipe

"Put lumps of quick-lime into a bucket of cold water and stir it about until it is all dissolved and mixed. It should be about as thick as cream. A pint of common varnish (which can be procured at a cabinetmaker's for a trifle) — will make it stick like paint. Instead of water to mix the lime with, skim milk (which must be perfectly sweet) will make the whitewash very white and smooth, and prevent it rubbing off easily — put on with a very long-handled brush made for the purpose. When it is quite dry, it must be gone over with a second coat, and if the wall is very dirty or has been colored with yellow ochre, a third coat may be necessary."[5]

Contemporary Whitewash

- Three to three-and-a-half parts hydrated lime
- One part liquitex
- Two parts water

A Superior, More Expensive Whitewash

- Two-and-a-half to two-and-a-third parts hydrated lime
- One-and-a-half parts liquitex
- One part water

Sand Paint

Cut stone was thought to be more elegant than wood as a cladding, and wood was often cut to resemble dressed stone. The use of sand paint as a coating further accentuated this illusion. A historic formula directs that, "All the wood work outside to be painted three good coats with first quality pure white lead and linseed oil paint & sanded twice at the time of putting on the last two coats and made to imitate Portland stone."[6] A contemporary alternative uses a compressor-operated glitter spray gun (the sort designed for painting cars). Fine sand is sprayed onto a wet base paint made from white lead and linseed oil tinted with burnt umber, burnt sienna and synthetic yellow ochre artists' oil paints. Tooling to resemble stone can begin after the sand has been sprayed on.

A Plaster Mix for Purists

Combine hydrated lime with water and let soak six to twelve hours. Add the lime to fine silica sand at a ratio of one-and-a-half to three parts sand to one part lime. Mix in one to two pounds of animal hair for each two to three cubic feet of plaster. A small amount of gypsum plaster will give added strength. Trowel on one thin coat over the lath. Two additional coats should be laid on as the base coat hardens. The initial setting time is approximately two weeks. Final curing requires six to eight weeks.

An Expedient Plaster Formula

Mix perlited gypsum plaster (Structolite) in a thirty-gallon trash can. Add jute or animal hair and water. Setting time, thirty to forty-five minutes for each coat. Final curing time, one to two days.

A Common Plaster Mix

Combine gypsum with an aggregate such as wood fibers, sand, perlite or vermiculite. Add water to a workable consistency.

Plaster for Use on Stone and Brick

- First two coats: two-and-a-half parts "riddled" lime (one part calcinated lime to three parts sand) to one part slaked or putty lime. Add seven pounds of animal hair for every ten cubic feet of mix. Use enough water to obtain a smooth mix.
- Finishing coat: one part sharp sand to one-and-a-half parts putty lime; water.

Plaster for Use on Metal Lath

- First two coats: two parts lime, one part Portland cement, nine parts sand, animal hair, water.
- Finishing coat: one to two parts white lime putty mixed with one to two parts gypsum plaster; sand (optional); water.

To Simulate Old Plaster

Drywall compound (durabond) should be mixed to a soupy consistency and then rolled onto new drywall (make sure the seams are staggered) or over badly spalled old plaster. Trowel it smooth with a twelve-inch mason's trowel or a rectangular plasterer's trowel. Let it dry slowly, because otherwise the material may crack. Any cracks that do develop may be repaired using the same material. This finish is cheap, easy to apply and relatively accurate in aesthetic quality and texture.

Fungicidal Wash for Plaster

A solution of sodium orthophenylphenate mixed and applied according to manufacturer's instructions.

Period Wallpaper Paste

Boil flour, water and a small quantity of alum until quite thick. Apply with a whitewash brush.

Homemade Distemper

- Whiting (approximately six-and-a-half pounds for a small room)
- Decorator's glue-size (approximately one pound per nine quarts of water)
- Dry, ground artist's pigment

Make up the glue-size following the directions on the container. It contains alum to check mold. Let the size cool until it has a jelly-like consistency. Reheat it by placing its container in a larger container of water over heat. When it is warm and runny, it is ready for use. Fill a small bucket half full with cold water and add whiting to a peak four inches above the water. Let it stand for one-and-a-half to two hours; stir. Dissolve the powder color in a small amount of cold water. Slowly add it to the whiting until the desired tint is achieved. Stir until the color is even, then add the warm glue-size and mix thoroughly. The consistency should be like thick paint. If the distemper begins to thicken further, place the container in a larger bucket of hot water. The mixture will not keep for more than a day or two. Distemper applied over plaster has a warm, deep, powdery look that can't be approximated using any other type of coating.[7]

Alternatives to Distemper

Over a mid-sheen oil-base undercoat, apply a thinned wash of tinted, flat oil-based paint.

A similar, more powdery effect can be gotten from using latex thinned with water and applied over a flat latex base coat.

More than one coat of these washes will give added depth, but second and third coats must be brushed on quickly to prevent dislodging the coat below.

Methods

All the methods listed here require care and attention to do them successfully. Some, like chemical cleaning of masonry, can involve a fair amount of risk for the novice. Be forewarned and take even more precautions than you think necessary. Still other methods, like enlarging a basement, sound easy enough on paper but involve strenuous work. Think about the amount of labor and time needed before beginning.

Even if you don't choose to undertake any of these procedures yourself, at least you will be aware of how they are done when the time comes to hire a contractor.

Estimating

When estimating the amount of recycled stone needed, first lay out or measure the outside wall (height and width), the inside wall (height and width) and the interior core. The total amount required is the cubic footage (length × width × height).

In a twenty-four-inch thick rubblestone wall, the building stones in both the interior and exterior parts will vary from eight inches to twelve inches thick, leaving a gap in the center from eight inches to practically nothing. The void must be filled with rubble and mortar. In my experience with new rubblestone walls, the core tends to be filled better today than in period construction, therefore requiring more material. Make sure you have enough rubble.

Abandoned rubblestone buildings are excellent sources of material. Cost may vary according to the area, the condition of the structure and the vendor. I recently purchased a thirty-five-foot by twenty-two-foot stone shell for less than $500. Six months earlier, I had been offered an entire building, smaller and in worse repair, for $12,000. Remember that the greatest cost is in transportation, and that old mortar must be removed before new construction can begin. Lime-based mortar is soft and removal should not be too difficult, requiring some chipping with hammer and chisel. A traditional method is to bury the stones in a leaf pile for the winter. Presumably, the acids in the decomposing leaves dissolves the mortar.

Brick recycling, although popular with pseudo-preservationists, scrap dealers and uptown individuals, has its pitfalls. In an original brick wall composed of more than one layer, only the exterior bricks will be long-wearing enough for use in exterior repairs or for building an addition. Interior bricks, known as salmon bricks, are soft, porous and unsuited for exterior walls. It is also imperative to know where recycled brick originally came from and the age of the building. Old brick may be machine-made in 1890 or soft, handmade brick from the 1700s, and the price per unit will vary accordingly.

In estimating brick requirements for repairs, the most accurate method is to actually count the number needed and allow five percent for waste and breakage. For additions or any kind of new construction, the Brick Institute of America offers a brick estimator for the paltry sum of sixty cents. The address is: 1750 Old Meadow Road, McLean, Virginia 22101.

Because of their weight, stone, brick and mortar materials are best trucked to your site by the supplier. Be sure to allow for the cost of gas and oil for the mortar mixer.

Estimates of repointing materials generally practised in new masonry construction are not applicable to period houses. The amount of mortar used will depend on the width of the original jointing, and the best estimating method for the novice is to buy a small quantity of hydrated lime, white Portland cement and sand and experiment yourself. To gauge costs, every wall should have a grid laid out over a drawing of it. In this way, more expensive problem areas can be priced and areas of relative ease in the chipping and repointing process identified.

Cleaning Masonry

There are three basic methods for cleaning masonry surfaces: mechanical, water and chemical. Mechanical cleaning, like sandblasting, uses abrasives to blast the dirt from the surface. These materials might be sand, slag, corn cobs or glass beads carried by a jet of water, air or a combination of the two. Although this method is fast and economical, it is difficult to control, detrimental to masonry surfaces generally, erases historic tooling marks, dulls edges and corners, creates excessive dust and may destroy vegetation. However, it is recommended for joint cleaning where cement staining has occurred and for surface dusting of some stone faces (limestone, granite, sandstone). Other mechanical systems, such as wire brushes, grinders and scalers, all cause scarring. Use copper wire bristles when brushing stone.

Water cleaning softens and rinses dirt from masonry surfaces. It can be done with low pressure, high pressure or by means of steam. The first two forms are reasonable in cost, but steam cleaning is expensive. Water cleaning can be accomplished easily by the layman, and all three types are relatively harmless to masonry and vegetation. However, prolonged water washing may lead to staining, and excessive moisture penetration may result in problems with interior plaster or trim. This type of cleaning cannot be done when freezing temperatures are likely. Steam cleaning is somewhat hazardous and requires complete safety equipment to prevent scalding the operator.

Chemical cleaning involves the application of a specific chemical for the specific job at hand which is then removed by means of brushes and water. It is relatively easy to control, will not mar surfaces and may be applied to isolated areas. There is a health hazard from fumes and caustic liquids. Overspray can damage surrounding buildings, vegetation and pets. Incorrect chemicals for the specific job may result in detrimental reactions with the joints and surfaces. Chemical cleaning is ideal for removing stains on masonry. Listed below are the proper chemicals for various types of stains:

Algae
Benzalkonium Chloride and Gloquat C. These are supplied as fifty percent solutions which must be diluted for use to 1 percent of the active compound; one gallon diluted to twenty-five gallons with tap water, preferably shortly before use. Application must be generous flood spray to give a coverage of about one hundred square feet per gallon. Inadequate treatment will give poor control and limited duration.

Lichens
Tributyltin oxide in aqueous formulation. This product is diluted with tap water and applied by generous flood spray.

Ivy
An ammonium sulphate paste should be painted onto the root, which will wither and die. If the roots of the ivy have penetrated the building, they should be cut at their points of entry and treated in a similar fashion. Remove stains with zinc or magnesium silico fluoride diluted on a one to forty ratio with water.

Oil and Grease
Carbon tetrachloride in a poultice (see below).

Iron and Steel
Oxalic acid and water (one pound oxalic acid to one gallon water). Spray on, brush, then hose off. Ammonium bifluoride increases the speed of the process. *Danger*: hydrofluoric acid is given off and may be dangerous to health.

Gutter Stains
One part sal ammoniac, four parts fullers earth or other binder, two parts household ammonia. Fix as a poultice and brush off.

Tar

Scrape, clean with toluene or benzine in a poultice. Flush with mild detergent.[8]

Poultices or leaching packs are very effective for stain removal. Mix fullers earth, diatomaceous earth or sawdust and water (or one of the chemicals mentioned above) into a wet, floury paste and apply directly to the stain. As the paste dries out, the stain will be drawn into it. The paste will then fall away, carrying most of the stain with it. The remainder may then be removed by using a stiff bristle brush.

Mechanical, water and chemical cleaning methods require highly specialized equipment and protective clothing. Talk to a building-cleaning specialist, explain what you are trying to accomplish and take his advice.

Enlarging a Basement

All too often the early house was built with less than a full basement, and because contemporary living frequently requires such an area for mechanical systems or work space, the enlargement of the basement becomes necessary. There is no easy way to excavate a basement in an already established house. The obvious problem is how to dig the earth and remove it. Removal is particularly difficult in a historic building without a basement entrance; some have only windows.

The cheapest method is also the most time-consuming and unpleasant. Start digging with a shovel and remove the earth by wheelbarrow through the entrance, or pail it out through a window. However, most rental outlets will have a small conveyor for hire. Dig a small void on the exterior side of the basement window to accept the conveyor end. A pickup truck bed can be positioned at the other end. In urban areas, where one of the problems is where to deposit the excavated earth, this solution is ideal, because the truck can simply be driven off to the nearest landfill site.

The actual excavation can take two forms: one will give you some additional area, and the second will result in a more or less full basement. Both procedures are essentially structural in nature, and an examination by an experienced consulting engineer would be money well spent.

In the first type of excavation, for every foot below the existing footings that you will be digging, come in two feet from the existing walls. This will reduce floor area, but your house will not fall down. When you have excavated to the desired depth, construct a series of forms around the walls of the excavation. The forms should protrude six to nine inches above where the original footing is (see illustration on page 173). Make them from half- to three-quarter-inch plywood, supported by stakes and cross-braced.

If you are wealthy, order enough reinforcing rod ($\frac{3}{8}$ to $\frac{1}{2}$-inch diameter) to form a grid for the floor inside the forms, using stovepipe wire to hold the grid together. If you are without funds, any metal objects will do: old bedsprings, bedside rails, bicycles, etc. It is now time to think about the concrete. The simplest way to do this efficiently is to call the concrete company and explain the type of work you are doing and the approximate dimensions. Be sure to specify how the mix is to get into the basement — through a window, via a wheelbarrow or whatever. They will design the mix and estimate the amount needed.

It is important to schedule the arrival of the concrete for early in the morning, because troweling will take six to twelve hours. Stage one is the placement of the mix, stage two is the initial leveling with a length of two-by-four and stage three is troweling and finishing.

When the mix arrives, be prepared to fill the forms first. You will be responsible for placing the concrete and making sure it is settled into place. Ram a pole up and down in the mix until all voids are completely filled. For large jobs, hire a concrete vibrator for the

A conveyor belt carries earth from the basement to the truck for easy disposal.

day. At this point, pour the floor onto a well-prepared base. The earth should have been well compacted beforehand and a four- to six-inch bed of crushed stone laid down and leveled. Spread a six-mil polyethylene vapor barrier over the stone; the concrete is laid directly over the vapor barrier. Be sure to allow for a sump hole if ground water will be a problem. Once the mix is spread over the entire floor area to a depth of about six inches, level it with the two-by-four. Let the concrete set up for about three to six hours, and then begin finishing with a trowel. On a small job, hand troweling will suffice, but a power trowel would be more efficient on larger areas. The floor should not be walked on for two or three days; complete set-up and curing will take four weeks. This form of extending the usable space is practical, relatively simple for the novice and cost-effective in the average preservation project.

This first type of excavation may not result in sufficient enlargement for all purposes. The second type will result in more space, but it requires a more complex and exacting type of excavation. Dig out the center area down to the desired depth as in the first method. Now select one area about three feet wide along the perimeter of the existing foundation. Dig this down to the new floor level. What you have is a void under part of the existing foundation that is the same thickness as the foundation and the same depth as the new floor. Excavate a few more of these voids at different spots along the foundation wall, build forms and fill them with concrete. Once the mix has set, remove the forms and repeat the procedure a number of times until the entire old foundation is resting on a new concrete foundation. As can be imagined, the number of voids excavated at any one time is limited by the degree of stability in the earth remaining under the old foundation. In an eighteen-inch-thick wall thirty feet long, it might be advisable to excavate only three non-consecutive voids at any one time (see illustration). Once the entire perimeter has been finished, a new floor may be poured and finished as described above.

Even if you do not wish to enlarge the basement area, it is often desirable to install a concrete slab to cover the original earth or nondescript floor. The established floor is probably already sufficiently compacted through age to offer a certain degree of stability and load bearing. After digging a sump pit if one is necessary, level the floor with four to six inches of crushed stone, then pour and level the floor as described above.

Rebuilding a Basement Entrance

The basement entrance was originally designed for convenience, but over the years it has probably become a major source of moisture penetration, heat loss and a host of related problems. The actual construction consisted of a break in the foundation, a wooden or masonry lintel above it, a rough frame lag bolted into the walls of the opening and a basement door hung on this frame. A set of stairs, usually stone, were built and encompassed by rubblestone walls approximately two feet thick. A storm door covered the entire structure.

Ground buildup around the storm door results in water penetration, which, in turn, causes failure of the rubblestone walls as footings get washed away. The basement door rots from the bottom, allowing beasts and climate to penetrate.

Although surface remedies in the form of repointing, door replacement and regrading may offer temporary relief, to ensure long-term, trouble-free use the entrance should be rebuilt. The first thing to do is to photograph or sketch the entrance from as many angles as possible. Remove the stones from the two walls and place them in two piles in close proximity to the work. Form another pile of the stones from the steps. Excavate to sufficient footing depth, paying attention to any weeping tiles that have been or will be installed. Construct forms around an area one foot wider and longer than the entire assembly occupied. Rent a compactor and run it over the earth until it is firm and even, bearing in mind that the level will have to allow for crushed stone and cement. Lay down a bed of crushed stone to a depth of four to six inches and over it place wire mesh reinforcement with six-by-six spacing and six-six gauge. Now pour a six-inch-deep concrete slab. The mortar in the original foundation is probably lime-based, and you should rebuild the entrance using similar material. Once the mortar has set, replace or reinstall the storm and basement doors.

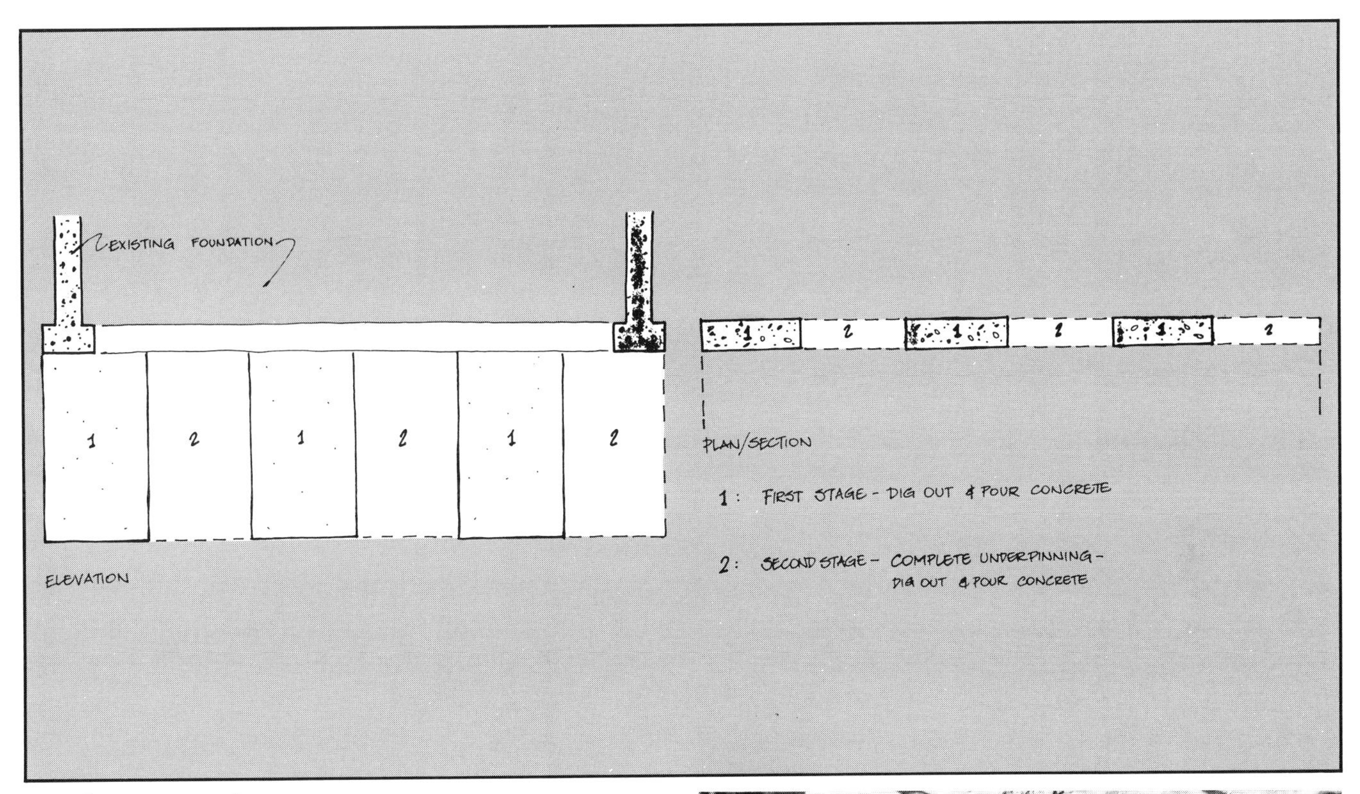

Underpinning a basement

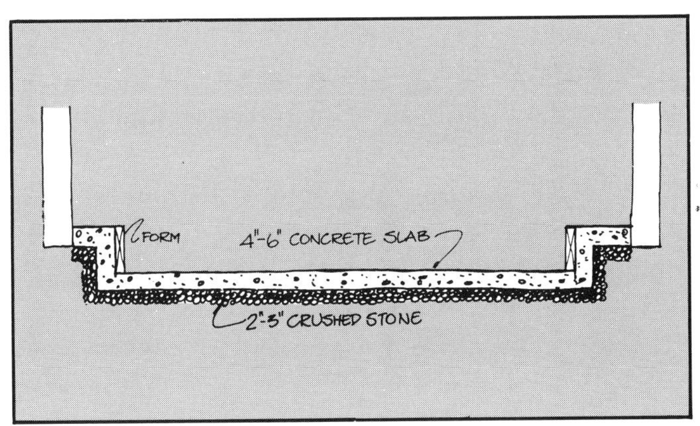

Cross-section of a concrete slab

This new basement entrance has been constructed using period materials and methods.

Notes

From the Old Land to the New *(pages 10 to 43)*

[1]Quoted in Sir Banister Fletcher, *A History of Architecture* (New York: Charles Scribner's Sons, 1975), p. 415.

[2]Quoted in C. F. Innocent, *Development of English Building Construction* (Devon, England: David & Charles, 1971), p. 121.

[3]Ibid., p. 129.

[4]J. L. Bishop, *A History of American Manufacture* (Philadelphia: Edward Young & Co., 1868), p. 220.

[5]Quoted in Hugh Morrison, *Early American Architecture from the First Colonial Settlements to the National Period* (New York: Oxford University Press, 1952), p. 69.

[6]Quoted in Bishop, *A History of American Manufacture*, p. 219.

[7]Gerald Finley, *In Praise of Older Buildings* (Kingston: Frontenac Historic Foundation, 1976), p. 16.

[8]Edward F. Bush, *The Builders of the Rideau Canal, 1826-32* (Ottawa: National Historic Parks & Sites Branch, Parks Canada, Dept. Of Indian & Northern Affairs, 1976), p. 12.

[9]Quoted in Olaf W. Shelgran, Jr., *et al.*, *Cobblestone Landmarks of New York State* (Syracuse, NY: Syracuse University Press, 1978), p. 19.

[10]J. C. Louden, *An Encyclopedia of Cottage, Farm and Village Architecture and Furniture* (London: Longman, Brown, Green and Longmans, 1846), p. 112.

[11]Ralph Edwards and L. G. G. Ramsey, *The Connoisseur's Complete Period Guide to the Houses, Decorations, Furnishings and Chattels of the Classic Periods* (New York: Bonanza Books, 1956), p. 527.

[12]Association for Preservation Technology Bulletin 7 (No. 4, 1975), p. 19.

[13]T. Ritchie, *Canada Builds* (Toronto: University of Toronto Press, 1967), p. 60.

Basic Materials (*pages 44 to 55*)

[1]Ritchie, *Canada Builds*, p. 209.

[2]William A. Parks, *Report on the Building and Ornamental Stones of Canada* (Ottawa: Government Printing Bureau, 1912) p. 129.

[3]Peter N. Moogk, *Building a House in New France* (Toronto: McClelland and Stewart, 1977), p. 92.

[4]Ibid., p. 92.

[5]Ibid., p. 92.

[6]Ibid., p. 94.

[7]Louden, *An Encyclopedia*, p. 602.

Inspection and Maintenance (*pages 56 to 81*)

[1]Louden, *An Encyclopedia*, p. 260.

[2]Ibid, p. 261.

[3]Ibid, p. 260.

[4]Norman R. Weiss, "Cleaning of Building Exteriors," *Technology and Conservation* 6 (Winter 1981).

[5]Francis Phipps, *Colonial Kitchens, Their Furnishings and Their Gardens* (New York: Hawthorn Books, 1972), p. 21.

Chimneys, Fireplaces and Stoves (*pages 102 to 121*)

[1]Mark Girouard, *Life in the English Country House* (New Haven: Yale University Press, 1978), p. 35.

[2]Ibid., p. 36.

[3]Moogk, *Building a House in New France*, p. 90.

[4]Arthur Channing Downs Jr., "Nostalgia for the Fireplace, 1851," *Magazine Antiques* 121 (February 1982), p. 495.

[5]One good source for reproduction firebacks and stove plates is The Country Iron Foundry, Box 600, Paoli, PA, 19301. Send $2.00 for a catalog.

[6]Jeanne Minhinnick, *At Home in Upper Canada* (Toronto: Clarke, Irwin and Company, 1970), p. 93.

Interior Finishes (*pages 122 to 143*)

[1]Probably the best single guide for painting techniques and recipes is *Paint Magic* by Jocasta Innes. Published in 1981 and distributed by Van Nostrand Reinhold (New York, Toronto), it is not necessarily a book on period decor but an excellent source for decorators in general.

Masonry in the Landscape (*pages 144 to 159*)

[1]Rudy J. Favretti, and Joy Putman, *Landscapes and Gardens for Historic Buildings* (Nashville, TN: American Association for State and Local History, 1978), p. 20.

[2]Ibid., p. 20.

Recycling Non-Domestic Buildings (*pages 160 to 165*)

[1]Phipps, *Colonial Kitchens*, p. 29.

Tools, Mixes and Methods (*pages 166 to 181*)

[1]Dan Cruickshank and Peter Wyld, *London, the Art of Georgian Building* (London: Architectural Press, 1975), p. 192.

[2]A. J. Downing, *The Architecture of Country Houses* (New York: Dover, 1969), p. 64.

[3]Cruickshank, *London, The Art of Georgian Building*, p. 192.

[4]Downing, *The Architecture of Country Houses*, p. 187.

[5]Jeanne Minhinnick, "Some Personal Observations on the Use of Paint in Early Ontario," *Association for Preservation Technology Bulletin* 7 (No. 2, 1975), p. 19.

[6]*Association for Preservation Technology Bulletin* 7 (No. 4, 1975), p. 18.

[7]Innes, *Paint Magic*, p. 48.

[8]A. Davey, *et al.*, *The Care and Conservation of Georgian Houses* (Edinburgh: Paul Harris Pub. with Edinburgh New Town Conservation Committee, 1978), pp. 73, 74, 76.

Glossary

Absorption rate The weight of water absorbed when a brick is partially immersed for one minute, usually expressed in either grams or ounces per minute.

Abutment The part of a structure that directly receives the pressure of an arch or beam.

Admixtures Materials added to mortar as water-repellent or coloring agents, or to retard or hasten setting.

Adobe brick Large, roughly molded, sun-dried clay brick of varying sizes.

Aggregate A hard, inert material in various size fragments, mixed with cementing material to form concrete, mortar or plaster.

Anchor A piece or assemblage, usually metal, used to attach building parts to masonry or masonry materials.

Angle brick Any brick shaped to an oblique angle to fit a salient corner.

Arch A curved compressive structural member, spanning openings or recesses; also built flat. *See also* Back arch; Jack arch; Relieving arch; Trimmer arch.

Arch brick A wedge-shaped brick for use in an arch, or an extremely hard-burned brick from an arch of a scove kiln.

Architectural terra cotta Hard-burned, glazed or unglazed clay building units, plain or ornamental, machine-extruded or hand-molded, and generally larger in size than brick or facing tile.

Ashlar Squared, hewn stone.

ASTM American Society for Testing and Materials.

Back arch A concealed arch carrying the backing of a wall where the exterior facing is carried by a lintel.

Back filling Rough masonry built behind a facing or between two faces; filling over the extrados of an arch; or brickwork in spaces between structural timbers, sometimes called *brick nogging*.

Backing The part of wall behind its face.

Bat A piece of brick.

Batter Recessing or sloping masonry back in successive courses; the opposite of *corbel*.

Bearing stone Stone supporting a load other than the units of which it is a part in the masonry work.

Bearing wall A wall or partition which supports weight.

Bed The top surface on which mortar is spread.

Bed joint The horizontal layer of mortar on which a masonry unit is laid.

Belt course A narrow, horizontal course of masonry, sometimes slightly projected, such as a window sill. Sometimes called *string course* or *still course*.

Blind bond Stone bond in which stone headers extend only halfway through the wall from face and backing; bonding in a wall that does not show on either side.

Blind header A concealed brick header in the interior of a wall, not showing on the face.

Bond Tying various parts of a masonry wall by lapping units one over another or by connecting with metal ties; patterns formed by exposed faces of units; or adhesion between mortar and masonry units or reinforcement.

Bond course A course consisting of units which overlap more than one wythe of masonry.

Breaking joints Any arrangement of masonry units that avoids continuous vertical joints.

Brick A solid masonry unit of clay or shale, formed into a rectangle while plastic and fired in a kiln. *See also* Adobe brick; Angle brick; Arch brick; Building brick; Clinker brick; Facing brick; Fire brick; Floor brick; Norman brick; Salmon brick; Soft-mud brick.

Building brick Brick for building purposes, not specially treated for texture or color. Formerly called *common brick*.

Buttering Placing mortar on a masonry unit with a trowel.

Caulking Filling cracks around window frames and expansion joints to make them weathertight.

Cavity wall A masonry wall built to provide an air space within the wall.

Cement A powder of alumina, silica, lime, iron oxide and magnesia burned together in a kiln and finely pulverized. When mixed with water, it forms a plastic mass that hardens by chemical combination.

Chimney cap Uppermost course or unit of a chimney stack, used to improve the draft.

Clinker brick A very hard-burned brick that is distorted or bloated because of nearly complete vitrification.

Closer The last brick or tile laid in a course. It may be whole or a portion of a unit. *See also* King closer.

Cobblestone Natural rounded stone, larger than a pebble and smaller than a boulder.

Column An ornamental or supporting pillar.

Composite wall Any bonded wall constructed of different masonry units.

Concrete Hard, strong building material made of cement, aggregate and water.

Coping The material or masonry units forming a cap or finish on top of a wall, chimney, etc. It protects masonry below from the penetration of water from above.

Corbel A shelf or ledge formed by projecting successive courses of masonry out from the face of the wall; the opposite of *batter*.

Cornice Top course of a wall, projecting horizontally.

Course One of the continuous horizontal layers of masonry units. *See also* Bond course; Damp course.

Curing Perfecting by chemical change. Having acquired its initial set and hardened, mortar is said to be in its curing stage.

Curtain wall A non-bearing wall. Built for the enclosure of a building; it is not supported at each storey.

Damp course A course or layer of impervious material that prevents the entrance of moisture from the ground or a lower course. Often called *damp check*. Also called water course.

Dog's tooth Bricks laid with the corners projecting from the wall face.

Dook A nailing strip inset into a wall to accept baseboards, chair rails or picture rails. Also called *furring strip*.

Dressing Working or finishing the face of a stone; also squaring a stone for ashlar.

Drip A projecting piece of material, shaped to throw off water and prevent its running down the face of a wall or other surface.

Dry stone Stone laid without mortar (e.g., a dry-stone wall).

Efflorescence A powder or stain sometimes found on the surface of masonry, resulting from water-soluble salts.

Facade Face of a building.

Facing The material that forms the finished surface of a wall.

Facing brick Brick made specifically for facing purposes, often treated to produce surface texture.

Fire brick Brick made of refractory ceramic material which will resist high temperatures.

Fire division wall Any wall subdividing a building to resist the spread of fire. It is not necessarily continuous through all stories to and above the roof.

Fire wall Any wall subdividing a building to resist the spread of fire and which extends from the foundation through the roof.

Flashing A thin, impervious material placed in mortar joints and through air spaces in masonry to prevent water penetration and/or provide water drainage.

Floor brick Smooth dense brick, highly resistant to wear, used for finished floors.

Foundation wall The part of a load-bearing wall below the level of the earth or below first floor beams and joists.

Frog A depression in the bed surface of a brick. Sometimes called a *panel*.

Furring A method of finishing the interior face of a masonry wall to provide space for insulation, prevent moisture penetration or to produce a level surface for finishing.

Gable Triangular upper part of a wall at the end of a roof; triangular hood over a window or door; triangular break in an eave.

Grout A cementitious component of high water-cement ratio, permitting it to be poured into spaces within masonry walls.

Header A masonry unit that overlaps two or more adjacent wythes of masonry to tie them together. Also called *bonder*. *See also* Blind header.

Hearth The floor of a fireplace.

Hydrated lime Quicklime to which sufficient water has been added to convert the oxides to hydroxides.

Jack arch An arch with horizontal or nearly horizontal upper and lower surfaces. Also called a *flat* or *straight arch*.

Jointer A tool used for joint finishing work or jointing.

Keene's cement Hard-finish gypsum plaster to which alum has been added.

King closer A brick cut diagonally to have one two-inch end and one full width end.

Lime Calcium carbonate; a binder. *See also* Hydrated lime.

Lime putty Hydrated lime in plastic form ready for addition to mortar.

Lintel Horizontal architectural member spanning an opening and carrying the load above it.

Mantel Beam, stone or arch supporting masonry above a fireplace; fireplace finish covering front and side of the chimney; also the ornamental shelf above a fireplace opening.

Masonry cement A mill-mixed mortar to which sand and water must be added.

Mortar A plastic mixture of cementitious materials, fine aggregate and water.

Norman brick A brick whose nominal dimensions are $2^2/_3'' \times 4'' \times 12''$.

Pargeting The process of applying one coat of cement mortar to masonry. Often spelled and/or pronounced *parging*.

Parapet wall That part of a wall entirely above the roof line.

Party wall A shared wall between adjoining buildings.

Pick and dip A method of laying brick whereby the bricklayer simultaneously picks up a brick with one hand and, with the other hand, enough mortar on a trowel to lay the brick. Sometimes called the *Eastern* or *New England method*.

Pilaster A projecting portion of a wall serving as a vertical column and/or beam.

Pitch A slope from zero height at one end to inches or feet at the other end (e.g., pitch or roof).

Pointing Troweling mortar into a joint after masonry units are laid.

Portland cement Hydraulic cement made of finely pulverized clinker; a mixture of argillaceous and calcareous materials; a binder.

Queen closer A cut brick having a nominal two-inch horizontal face dimension.

Quoin A projecting right-angle masonry corner.

Racking A method entailing stepping back successive courses of masonry.

Relieving arch One built over a lintel, flat arch or smaller arch to divert loads, relieving the lower member from excessive strain. Also known as a *discharging* or *safety* arch.

Return Any surface turned back from the face of a principal surface.

Reveal That portion of a jamb or recess which is visible from the face of a wall back to the frame placed between jambs.

Rough cast A common nineteenth-century stucco or parging, which had small stones applied to it before it set.

Rough-pointing Where a mortar joint is smeared with the trowel, presenting a very sloppy and rustic mess.

Salmon brick Relatively soft, under-burned brick, so named because of color. Sometimes called *chuff* or *place brick*.

Sandblasting Engraving, cutting or cleaning with a high velocity stream of sand forcibly projected by air or steam.

Shoved joints Vertical joints filled by shoving a brick against the next brick when it is being laid in a bed of mortar.

Sill Course or courses of masonry with an inclined face against which *voussoirs* of an arch abut; a unit placed to receive pressure.

Slaking Treating lime with water, causing it to heat and crumble.

Smoke chamber That part of a fireplace between the top of the throat and the bottom of the flue.

Smoke shelf The baffle in a flue designed to prevent downdrafts.

Soffit The underside of a part of a ceiling, overhang or cornice.

Soft-burned Clay products that have been fired at low temperatures, producing relatively high absorptions and low compressive strengths.

Soft-mud brick Brick produced by molding relatively wet clay; often a hand process. When the insides of molds are sanded to prevent sticking of clay, the product is sand-struck brick. When molds are wetted to prevent sticking, the product is *water-struck brick*.

Soldier A stretcher set on end with the face showing on the wall surface.

Solid masonry wall A wall built of solid masonry units with joints completely filled with mortar.

Spall A small fragment removed from the face of a masonry unit by a blow or by action of the elements.

Still course *See* Belt course.

Stretcher A masonry unit laid with its greatest dimension horizontal and its face parallel to the wall face.

String course *See* Belt course.

Struck joint Any mortar joint which has been finished with a trowel.

Template Gauge or form used as a pattern or copy.

Throat Opening in the fireplace from the top of the firebox to the smoke chamber.

Tiles Hollow masonry building units composed of burned clay, shale, fire clay or mixtures thereof. Also called *structural clay*.

Tooling Compression and shaping the face of a mortar joint with a special tool other than a trowel.

Trimmer arch An arch, usually a low rise arch of brick, used for supporting a fireplace hearth.

Tuckpointing Filling in cut-out or defective mortar joints with fresh mortar.

Veneer A single wythe of masonry for facing purposes, not structurally bonded.

Wainscot Lower three or four feet of an interior wall when finished differently from the rest of the wall.

Wall tile A bonder or metal piece that connects wythes of masonry to each other or to other materials.

Water table A line of molded bricks or stone set out from the facade of a building where a visible basement joins the main house.

Weep holes Openings placed in mortar joints of facing material at the level of flashing, to permit the escape of moisture.

Wythe Each continuous vertical section of masonry, one unit in thickness; the thickness of masonry separating flues in a chimney. Also called *withe* or *tier*.

Veneered wall A wall having a masonry facing attached to the backing but not bonded to exert common action under load.

Water course *See* Damp course.

Bibliography

Ashurst, John, and Francis G. Dimes. *Stone Building: Its Use and Potential Today*. New York: Architectural Press, 1977.

Blake, Verschoyle Benson, and Ralph Greenhill. *Rural Ontario*. Toronto: University of Toronto Press, 1969.

Brown, George W. *Building the Canadian Nation*. Toronto: J. M. Dent & Sons (Canada), 1942.

Bureau of Naval Personnel. *Basic Construction Techniques for Houses and Small Buildings Simply Explained*. New York: Dover Publications, 1972.

Bush, Edward F. "Builders of the Rideau Canal." Manuscript Report Number 185. Ottawa: Parks Canada, 1976.

Ching, Francis D. K. *Building Construction Illustrated*. New York: Van Nostrand Reinhold, 1975.

Cruickshank, Dan, and Peter Wyld. *London: The Art of Georgian Building*. London: Architectural Press, 1975.

Davey, Andy, *et al. The Care and Conservation of Georgian Houses*. Edinburgh: Paul Harris Publishing, 1978.

Davison, J. I. *Masonry Mortar*. Ottawa: National Research Council of Canada, 1974.

Dezettel, Louis M. *Masons and Builders Library: Bricklaying, Plastering, Rock Masonry, Clay Tile.* Indianapolis: Theodore Audel & Co., 1972.

Downing, A. J. *The Architecture of Country Houses*. New York: Dover Publications, 1969.

Duncan, S. Blackwell. *The Complete Book of Outdoor Masonry*. Blue Ridge Summit, PA: Tab Books, 1978.

Edward, Ralph, and L.G.G. Ramsey. *The Connoisseur's Complete Period Guides to the Houses, Decorations, Furnishing & Chattels of Classic Periods*. New York: Bonanza Books, 1956.

Favretti, Rudy J., and Joy Putman. *Landscapes and Gardens for Historic Buildings*. Nashville: American Association for State and Local History, 1978.

Finley, Gerald. *In Praise of Older Buildings*. Kingston: Frontenac Historic Foundation, 1976.

Fletcher, Sir Banister. *A History of Architecture*. New York: Charles Scribner's Sons, 1975.

Girouard, Mark. *Life in the English Country House*. New Haven: Yale University Press, 1978.

Grimmer, Anne E. "Dangers of Abrasive Cleaning to Historic Buildings." Preservation Briefs: 6. Washington, DC: U.S. Department of the Interior, 1979.

Gross, James G., and Harry C. Plummer. *Principles of Clay Masonry Construction: Students Manual*. Mollen, VA: Brick Institute of America, 1973.

Handegord, G. O., and K. K. Karpati. *Joint Movement and Sealant Selection*. Ottawa: National Research Council of Canada, 1973.

Hearn, John. *The Canadian Old House Catalogue*. Toronto: Van Nostrand Reinhold, 1980.

Humphreys, Barbara A., and Meredith Sykes. *The Buildings of Canada: A Guide to Pre-20th Century Styles in Houses, Churches and Other Structures*. Montreal: Parks Canada, 1974.

Hutchins, Nigel. *Restoring Old Houses*. Toronto: Van Nostrand Reinhold 1980.

Innes, Jocasta. *Paint Magic*. Toronto and New York: Van Nostrand Reinhold, 1981.

Innocent, C. F. *Development of English Building Construction*. Devon, England: David & Charles, 1971.

Insall, Donald. *The Care of Old Buildings Today*. London: The Architectural Press, 1972.

Jordan, R. Fourneaux. *A Concise History of Western Architecture*. London: Thames and Hudson, 1971.

Lenczner, David. *Elements of Loadbearing Brickwork*. Oxford: Pergamon Press, 1972.

Lessard, Michel, and Hugette Marquis. *Encyclopedie de la Maison Québecoise*. Ottawa: Les Editions de l'Homme Ltée, 1972.

Lockwood, Glenn J. *Montague: A Social History of an Irish Ontario Township*. Smiths Falls, Ont.: Township of Montague, 1980.

Loudon, J. C. *An Encyclopedia of Cottage, Farm & Villa Architecture & Furniture*. London: Longman, Brown, Green and Longmans, 1846.

Mack, Robert C., De Teel Pat Tiller, and James S. Askins. "Repointing Mortar Joints in Historic Brick Buildings." Preservation Briefs: 2. Washington, DC: US Department of the Interior, 1979.

MacRae, Marion, and Anthony Adamson. *The Ancestral Roof*. Toronto: Clarke, Irwin & Company, 1963.

McKee, J. *Introduction into Early American Masonry, Stone, Brick, Mortar, and Plaster*. Washington, DC: National Trust for Historic Preservation, 1973.

Minhinnick, Jeanne. *At Home in Upper Canada*. Toronto: Clarke, Irwin & Company, 1970.

Moogk, Peter N. *Building a House in New France*. Toronto: McClelland and Stewart, 1977.

Morrison, Hugh. *Early American Architecture from the First Colonial Settlements to the National Period*. New York: Oxford University Press, 1952.

Nicholson, Peter. *The Mechanic's Companion*. Philadelphia: F. Dell, 1852.

Nickey, J. M. *The Stoneworker's Bible*. Blue Ridge Summit, PA: Tab Books, 1979.

The Old House Journal. New York: The Old House Journal Corporation.

Peterson, Charles E. *Building Early America: Contributions Toward the History of a Great Industry*. Radnor, PA: Chilton Book Company, 1973.

Phipps, Francis. *Colonial Kitchens, Their Furnishings and Their Gardens*. New York: Hawthorn Books, 1972.

Phillips, Morgan W. *The Morse-Libby Mansion*. Washington, DC: US Department of the Interior, 1977.

Rawson, Marion. *Sing Old House*. New York: E. P. Dutton & Co., 1934.

Rempel, John I. *Building with Wood*. Toronto: University of Toronto Press, 1980.

Ritchie, T. *Canada Builds*. Toronto: University of Toronto Press, 1967.

———. *Silicone Water-Repellants for Masonry*. Ottawa: National Research Council of Canada, 1974.

Rothery, Sean. *Everyday Buildings of Ireland*. Dublin: Department of Agriculture, College of Technology, 1975.

Service, Alastair. *Edwardian Architecture and Its Origins*. Hampshire, England: Architectural Press, 1975.

Shelgran, Olaf W. Jr., Cary Lattin, and R. W. Frasch. *Cobblestone Landmarks of New York State*. Syracuse, NY: Syracuse University Press, 1978.

Slater, Gerald A. *Stone Preservatives: Methods of Laboratory Testing and Preliminary Performance Criteria*. Washington, DC: W. S. Gort Printing Office, 1977.

Sloane, Eric. *A Museum of Early American Tools*. New York: Funk & Wagnalls, 1964.

Stockham, Peter. *Little Book of Early American Crafts and Trades*. New York: Dover Publications, 1976.

Tallmadge, Thomas E. *The Story of Architecture in America*. New York: W. W. Norton & Company, 1927.

Vivian, John. *Building Stone Walls*. Charlotte VT: Garden-way Publishers, 1976.

Wagner, Willis H. *Modern Carpentry*. South Holland, IL: The Goodheart-Willcox Co., 1976.

Weaver, Martin. *Tips on Home Maintenance in Canada*. Ottawa: The Heritage Canada Foundation, 1980.

Wilson, P. Roy. *The Beautiful Old Houses of Quebec*. Toronto: University of Toronto Press, 1975.

Index